Good Owner's Clock Guide
and
Clock Logbook

Good Owner's Clock Guide and Clock Logbook

JOHN MOORHOUSE

Foreword by Eliot Isaacs
Curator, British Horological Institute

N.A.G. Press
an imprint of Robert Hale • London

Good Owner's Clock Guide

© *John Moorhouse 1999*
First published in Great Britain 1999

ISBN 0 7198 0280 6

Robert Hale Limited
Clerkenwell House
Clerkenwell Green
London EC1R 0HT

Good Owner's Clock Logbook
published in a separate volume

ISBN 0 7198 0290 3

Typeset in 10/15 Sabon by
Derek Doyle & Associates, Mold, Flintshire.
Printed and bound by
WBC Book Manufacturers Limited, Bridgend

Good Owner's Clock Guide

Contents

Foreword

Museum curators have long realized the importance of maintaining comprehensive documentation for all the artefacts in their possession. As well as keeping records of how the object came to enter the collection, it is regarded as essential to keep track of repairs and conservation so that future generations can appreciate both the original condition of the item and any changes which have taken place whilst it has been in the care of the museum. Curators are also fully aware of what is necessary to maintain the condition of their collection in perpetuity, free from inappropriate repair or neglect.

Regrettably, the same concern or awareness is seldom shown by owners or private collectors. Clocks in particular, can undergo many changes, some of a fundamental nature, whilst in the hands of a succession of owners. Some arise from a lack of care; some are carried out in a praiseworthy attempt to keep the clock operating; others are made in an attempt to maximize the profit on a resale. Most horologists have their stories of escapements that appear to be the wrong period, or dials that obviously were never made for the particular movement.

Unfortunately the average owner or collector, whilst appreciating the artistic beauty, and perhaps the value of the clock in his or her care, often has insufficient knowledge to know when maintenance is required or to recognize when simple changes have been made, let alone be able to describe any repairs or other alterations in sufficient detail to be of much help to future owners. More importantly, perhaps, if a valuable clock has been stolen, the owner is frequently unable to provide more than the vaguest of details of its appearance to assist in its recovery.

These shortcomings could be avoided if the owner had a basic education in care and custodianship and had kept comprehensive records throughout the life of the clock, and had passed them from one owner to another.

This book sets out to remedy this state of affairs, and is long overdue. For the first time, the author has provided a standardized format in which to record the complete history of a clock, which can be used to prove ownership, as well as being able to assist future conservators. It has often been said that none of us owns a clock – we are merely temporary custodians. Now, at last, we are able to discharge our duties of custodianship in a responsible manner.

Eliot Isaacs MSc FBHI
Curator, British Horological Institute 1990–8

Acknowledgements

The author would like to record his sincere thanks to those who have provided assistance, advice and encouragement to bring this project to fruition: Dr T. Treffry, Editor of the *BHI Journal*, for his technical comments, Mrs Helen Bartlett, Secretary of the British Horological Institute and Mrs Jill Hadfield for their valuable advice regarding publication; Mr John Manasseh, the Hatton Garden silversmith and antique silver dealer, for his support and encouragement; Mr John Thompson for his advice and skills with the photographic prints; Mr David Penney for his excellent training and guide in the skills of horological illustration; Mr John Hale and his team at Robert Hale Ltd for their professional advice in the preparation of the book for publication and sale; Mr Paul Thurlby, the gifted tutor and Course Director at the University of Central England for providing an outstanding horological education, as well as technical comments on the text; customers and friends who kindly agreed to allow their clocks to be photographed for use in educating others; and his wife Janice, a jewel in the world, for general editing and reviewing.

Note: The photographs were taken by the author with the assistance of his wife. Illustrations by the author.

Introduction

The latter half of the twentieth century has seen an increasing interest in, and appreciation of, antique and collectable clocks. Whether an English longcase, French mantel, American wall, Black Forest alarm or fine-bracket clock, they are all part of the history of industry and its development in the Western world and form a horological heritage of which many countries are proud. A clock may be hand-made or mass-produced, a unique item made to exacting standards of construction or performance, or produced in large numbers to meet popular demand. In recent years, a new level of appreciation has had a number of effects. It has initiated large increases in real values – over a period of thirty years the price of an English longcase clock has risen thirty-fold. It has resulted in recent years in a noticeable reduction in the condition and grade of clocks for sale at auction, thereby reducing the opportunity to acquire a quality item. It has spurred further research into existing and new areas of horology, leading to more publications for those interested in restoring, making or collecting clocks.

Those who are collectors, or have a technical interest or awareness are therefore fully provided for. Ordinary clock owners or potential owners however, are in a less fortunate position. Owners have a number of needs: to be able to care for their cherished and valuable clock properly, to keep it safe from harm or neglect and to keep it running continuously with good timekeeping, whilst enjoying fully the benefits of ownership. Potential owners, spurred by a desire to own an attractive and desirable clock, needs to be fully aware of what is on offer in the current market place in terms of condition, quality and originality. Good clocks are always fully appreciated – their quality was recognized when they were first made and still is. Bargains may be available, but they are difficult to identify and obtain. On the other hand, duds are very common, either being severely damaged, badly distressed or assembled from parts of other clocks. Particular care needs to be taken so that a judicious selection can be made.

The *Good Owner's Clock Guide* aims to provide for the needs of both the owner and the potential owner by offering sound information and advice in all relevant areas, presented in a simple, non-technical way. It is applicable to all types of clock, but is primarily concerned with spring- or weight-driven clocks, since these represent the large majority of those in use. The information will ensure that the owner is confident of taking the correct action and will know when expert advice should be sought.

It has to be immediately acknowledged that to cover every type of clock is an almost impossible task. Fortunately, there are many common features in clocks and it is possible to draw examples from the most frequently found types in the knowledge that the information will be directly applicable to many other kinds: Cross generalizations are inevitable when attempting so ambitious a task, but the reader should not be deterred by them. Where a greater depth of information is required there are many sources of reference readily available. Where the information presented appears either trivial or obvious to those with prior knowledge, the author begs their patience and understanding.

When caring for a clock, it is beneficial for the owner to keep records of a number of aspects. Details of maintenance and repairs will be an important aid to future repairers. Records of valuation for insurance purposes will be useful in case of theft or damage. Full details of the clock movement and case are desirable to assist in its recovery if it is stolen. This will also be essential to prove ownership, and will be useful when restoring damage. Information on the history and ownership of the clock will be of interest to future owners. Records of timekeeping and any adjustments will be an asset in ensuring good performance. The *Good Owner's Clock Logbook* aims to meet all of these needs within one volume, allowing owners to take better care of their possessions.

The comprehensive record will also be of benefit when a clock is sold. It will prove the authenticity and value of the item and show that the previous owners were concerned about the well-being of the clock. The *Good Owner's Clock Logbook* will form a permanent source of reference for the owner and will hopefully accompany the clock through many future generations.

Colour Illustrations

17

(ix) A movement attached to a clock by Charles Oldham of Southam showing unused holes in the front plate of the thirty-hour movement

(x) A good example of a completely original fitting of an eight-day fusee movement to the dial in an office dial clock

between pages 176 & 177

(xi) The rear of a brass dial showing the hammered surface and flaws in the cast brass with no signs of change from originality

(xii) Part of the door and trunk of an early nineteenth-century Yorkshire longcase with reeded columns and inlaid cross-banding and both pictorial and geometrical inlays

(xiii)A very good quality brass dial, eight-day striking and chiming clock by Ellicott of London

(xiv)An unrestored painted dial with very little evidence of the original black lettering below the date aperture but which retains an impression of the maker's name and origin

Line Illustrations

1 Care and Maintenance

There are large numbers of antique and collectable clocks in private ownership, a majority of which will be in continuous operation. A clock is often a source of considerable pleasure for the owner, especially when it has been passed down within the family for many generations. It is a complicated and delicate instrument, and many are of high value and of considerable historical interest. A proper level of care and maintenance is essential if a clock is to give its owner a satisfactory level of service. Unfortunately, the owner will not normally have any special expertise in the care of such a delicate instrument. The clock may be neglected or inappropriately cared for.

This guide provides basic information and advice for owners on topics such as transport, storage, installation and simple adjustments. No attempt is made to provide a detailed horological knowledge, just a complete understanding of all the problems which can arise in everyday use and how either to avoid or to rectify them. A majority of the potential problems are well within the capabilities of any owner to resolve without in any way risking damage to the clock. On the contrary, it is intended that the owner will be able to avoid many of the common occurrences which can lead to damage through unintentional ignorance or neglect.

Care and basic maintenance are well within the scope of the lay person, but proper cleaning, lubrication, repair and restoration should be left to those with the necessary expert knowledge. Proper professional care is most important, since the lack of appropriate maintenance can affect the timekeeping, appearance, mechanical condition and value of a clock. Such care will ensure that the pleasures of ownership are available for many generations to come.

Transportation

Transporting a clock with the pendulum in its normal operating position can cause serious damage to the means of supporting the pendulum, and to the escapement of the clock. Whenever a clock is moved the pendulum should therefore be carefully lifted off its support (the back cock), lowered down until it is free of the movement, and placed in a safe position. It is usually suspended from a thin steel strip or a silk thread, and care should be taken to avoid bending the steel strip, if one is used, or the pendulum will not swing correctly.

In a mantel or bracket clock there is not usually enough space to lower the pendulum. It should therefore be lifted outwards once it is detached from the crutch, taking care to avoid the hammer and the bell or gong. It may be necessary to remove the bell or gong by loosening its screw. To avoid a silk thread becoming tangled, use a small weight such as a paper clip or rubber band to retain it to a convenient part of the movement.

In some good-quality mantel clocks it is not necessary to remove the pendulum; it can be locked safely in position, using a special fitting – either a screw which is 'parked' in the centre or the side of the clock backplate or a slide mounted onto the wooden support for the movement.

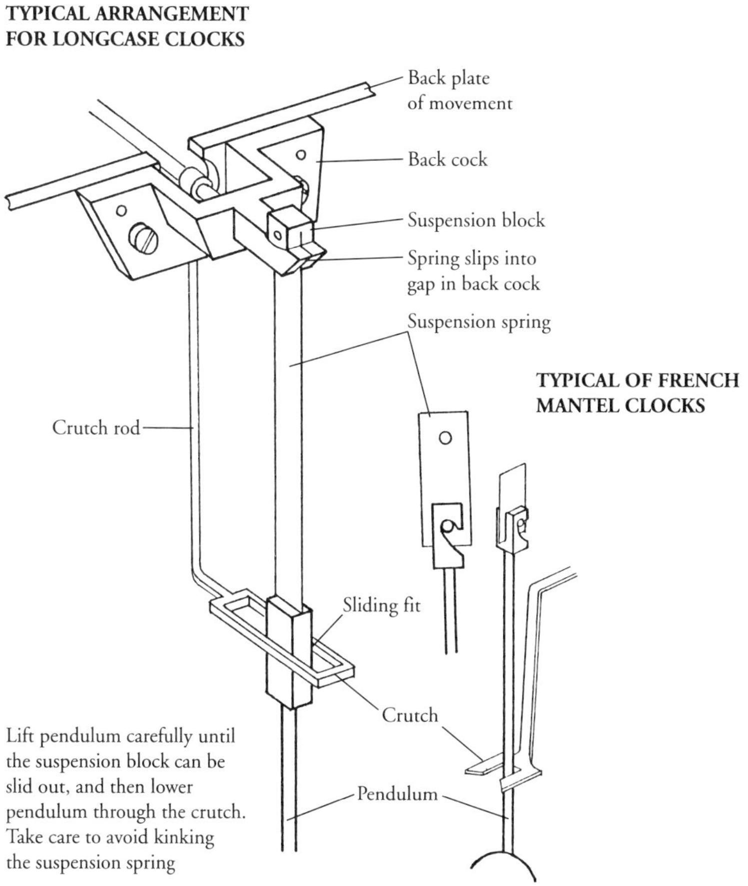

**TYPICAL ARRANGEMENT
FOR LONGCASE CLOCKS**

Back plate
of movement

Back cock

Suspension block

Spring slips into
gap in back cock

Suspension spring

**TYPICAL OF FRENCH
MANTEL CLOCKS**

Crutch rod

Sliding fit

Crutch

Lift pendulum carefully until
the suspension block can be
slid out, and then lower
pendulum through the crutch.
Take care to avoid kinking
the suspension spring

Pendulum

Fig. 1a   Attaching or removing the pendulum

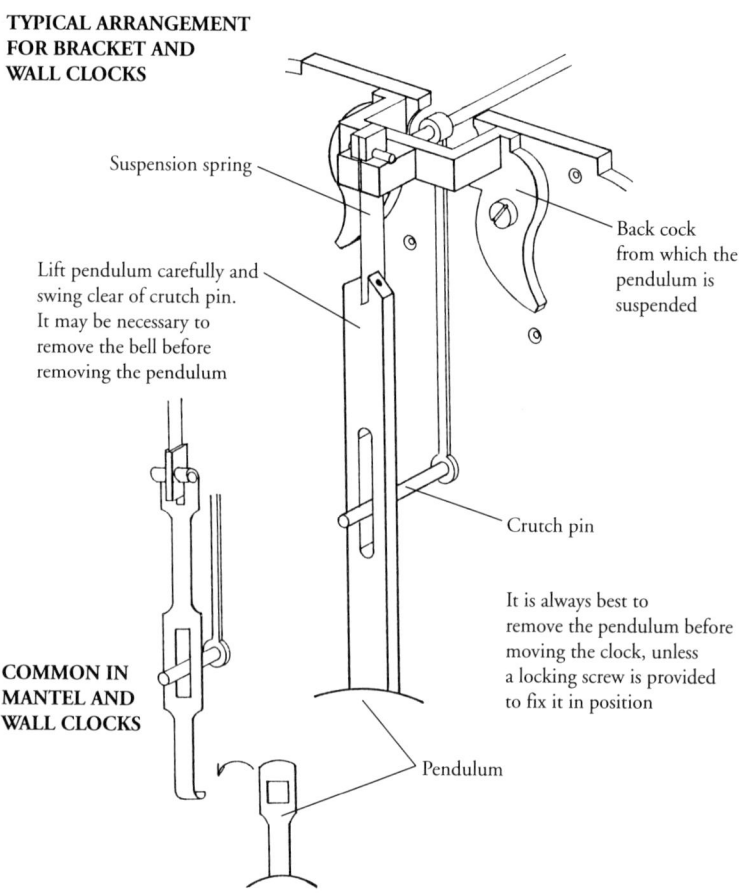

TYPICAL ARRANGEMENT FOR BRACKET AND WALL CLOCKS

Suspension spring

Back cock from which the pendulum is suspended

Lift pendulum carefully and swing clear of crutch pin. It may be necessary to remove the bell before removing the pendulum

Crutch pin

It is always best to remove the pendulum before moving the clock, unless a locking screw is provided to fix it in position

COMMON IN MANTEL AND WALL CLOCKS

Pendulum

Fig. 1b   Attaching or removing the pendulum

Some pendulums can be very heavy, particularly those containing mercury in a steel or glass jar. Great care needs to be taken to avoid any damage or spillage. Care also needs to be taken when removing a pendulum that the adjustment (rating) nut, which is normally situated below the pendulum, is not moved, as this will upset the rate of the clock.

When moving a small clock by hand from room to room, it is possible, with care, to transfer it without removing the pendulum. After stopping the pendulum the clock is tilted slowly forwards until the pendulum rests against the movement. This prevents it from swinging about or becoming detached from the crutch.

It is beneficial to transport a clock with the movement in the normal operating position – i.e. vertical – provided that it is securely fixed into the case. This will involve checking not only that the movement is attached to its supporting bracket, shelf or seatboard, but also that these are firmly attached to the case; if they are not, it may be advisable to remove the movement before moving the case.

Longcase clocks are an exception. It is almost always necessary to remove the movement before moving the clock to reduce the weight. The first step is to completely remove the hood, which usually slides forwards. This can be a tricky operation for those with no experience. A slight lift of the hood, whilst applying gentle pressure at the rear is usually necessary. Occasionally a knee resting against the front of the trunk is necessary to maintain the safety of the case. If the weights and pendulum are in position the security of the movement will be guaranteed. If they are not there is a risk of the movement falling out of the case. Normally, the movement will be held down to the seatboard with bolts or hooks and the seatboard will be fixed to the sides of the case (the cheeks) using screws or nails, but it is quite common for the movement and seatboard to be kept in position solely by the weights and pendulum. An extra pair of hands is highly desirable.

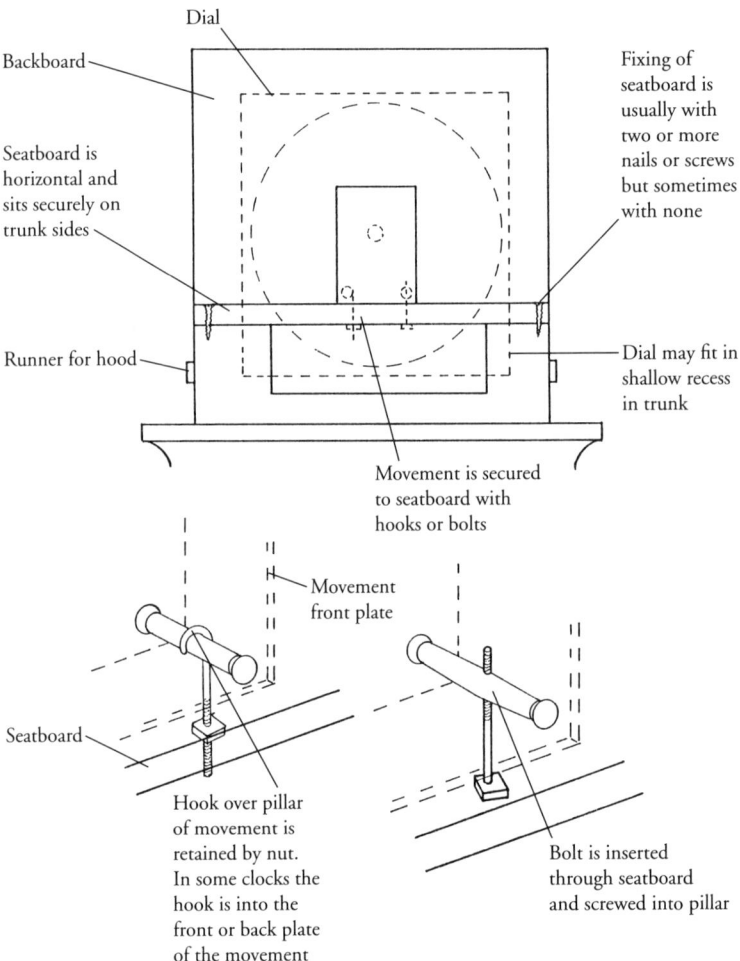

Fig. 2   Fitting of a movement into a longcase clock

Having removed the hood it is necessary to make certain the movement is safely secured with no risk of it falling out of the case. The weights can then be carefully removed. Each should be marked to ensure that they are correctly reinstalled on their respective pulleys. Steps will need to be taken to prevent the ropes, chain or gut becoming tangled and loose on their winding barrels. One way is to tie them loosely together in a large knot or use some light cord to tie them together neatly; another is to use masking tape around the gut on the barrel. The pendulum can now be carefully lifted off, threaded through the crutch and placed in a safe position. For trouble-free transport a pendulum can be strapped to a piece of wood slightly longer than itself. This will reduce the risk of bending and protect the thin steel suspension spring.

If the movement does have to be removed it should be placed in a vertical position in a strong cardboard box, packed around with crumpled paper or other suitable packing material. It is necessary in spring-driven clocks to avoid the striking or chiming trains trying to operate on the hour or quarter during transport or storage. This can be done by placing a folded piece of paper between the crutch and the backplate of the movement and is necessary to prevent the escapement operating during transit. Alternatively, the clock can be allowed to wind down completely.

Care also needs to be taken to prevent damage to the dial and hands. Avoid resting the movement on its dial, bezel or glass, or on the rear of the movement. Care should be taken to avoid bending the crutch.

Wall clocks are normally provided with front, bottom or side access doors to allow the movement to be viewed and the pendulum and any weights to be removed. In dial clocks, the main part of the case is attached to the circular frame by stout wooden pegs inserted at the sides, the movement and dial being attached to the circular frame. It is important to ensure that these pegs are secure before removing such a clock from the wall.

Clocks can be very heavy and special care needs to be taken when lifting them. It is always best to lift them from underneath. The case may not have the same strength as when it was first made, joints in marble and slate cases can be fragile, old wood may be brittle, and the handles provided on bracket clocks should not be trusted – some types of handle were fitted only as decoration. It is advisable to ensure that all doors are carefully locked or tied and the keys removed when moving any case, and that the glass in the door, case or bezel is adequately secure for travelling.

Installation

When choosing a location for the clock, consideration needs to be given to those locations which offer safety from disturbance, dust, damp, sources of heat and direct sunlight. The ideal conditions for a clock are 15–20 degrees centigrade and 50–60 per cent relative humidity. Before coming to a final decision on the location it is also wise to test whether the sound of the striking can cause a nuisance.

Longcase and mantel clocks should be placed on a firm and flat surface to prevent rocking or swaying of the case under the influence of the pendulum. In longcase clocks a large proportion of the weight is at the top, particularly when they are fully wound, and the case can therefore be unstable and prone to vibration. It is desirable to screw the clock to the wall. This will guarantee that it will not fall, move or be pulled over. Some rigid spacers may be required between the wall and the clock to ensure that it is upright. This is usually necessary when the base of the clock is against a skirting board.

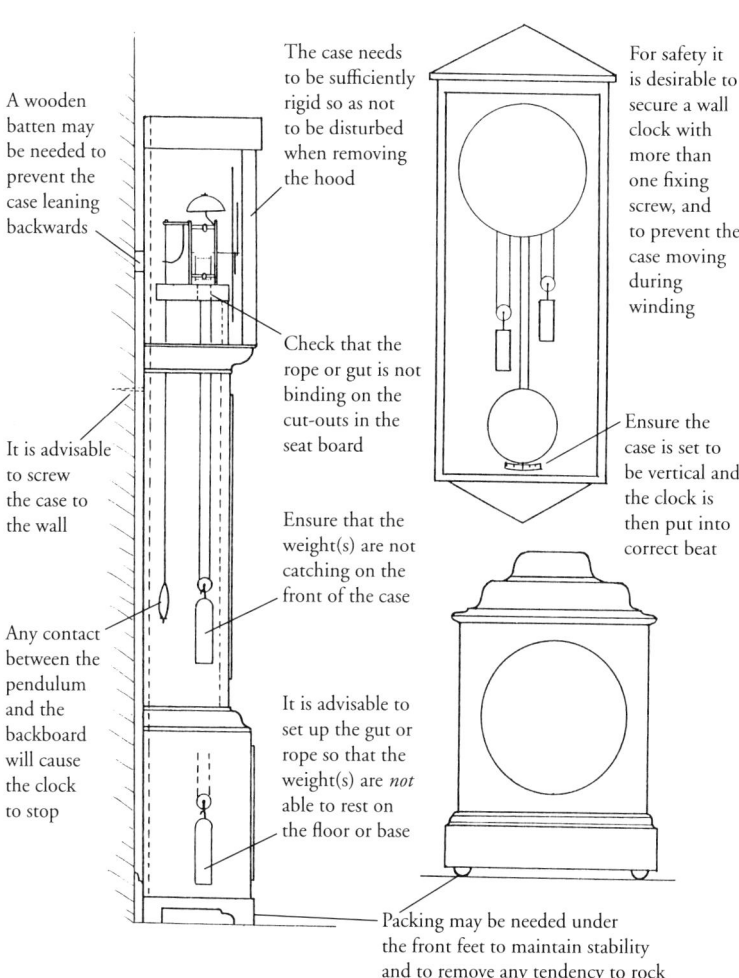

A wooden batten may be needed to prevent the case leaning backwards

The case needs to be sufficiently rigid so as not to be disturbed when removing the hood

For safety it is desirable to secure a wall clock with more than one fixing screw, and to prevent the case moving during winding

Check that the rope or gut is not binding on the cut-outs in the seat board

It is advisable to screw the case to the wall

Ensure the case is set to be vertical and the clock is then put into correct beat

Ensure that the weight(s) are not catching on the front of the case

Any contact between the pendulum and the backboard will cause the clock to stop

It is advisable to set up the gut or rope so that the weight(s) are *not* able to rest on the floor or base

Packing may be needed under the front feet to maintain stability and to remove any tendency to rock

Fig. 3   Correct installation of a case

Wall clocks should be firmly attached to the wall using screws or brackets which are guaranteed to be secure over a prolonged period of use; nails can rapidly work loose in plaster walls. Some means of preventing the clock moving from side to side during winding is desirable. This is when the greatest loads are imposed. An additional factor is the extra sideways load imposed by the weight of the open bezel and glass. It is beneficial if two separate means of attaching the case to the wall are used. Spikes attached to the cases of wall clocks are sometimes provided to help stabilize the case.

American shelf clocks are often shallow in depth and can be unstable when the door is opened. Fixing to the wall is advisable, particularly for weight driven clocks.

The case of a clock should be vertical. When properly restored this will ensure the clock will have an even tick, i.e. be in beat. This is when there are equal intervals between successive ticks. It is *most important* that the clock should have an even tick, in order for the power to be transferred correctly to the pendulum, and for the clock to operate for the full going period and give good timekeeping. An uneven tick, i.e. one that is out of beat, is a very common reason why a clock in otherwise good order fails to operate. It is essential for the owner either to be able to set the beat correctly or have it done by someone with the necessary experience. It is easy to carry out but it is possible to cause serious damage by taking inappropriate action. Detailed guidance is therefore provided below to allow the owner to judge whether an adjustment is necessary and to assess whether they would be able and willing to do it themselves.

How to Adjust the Beat

A crude method of bringing the clock into beat is to lift up one side of the case and insert small temporary chocks (both left and right sides will need to be tried to find which way the adjustment needs to be made) until the tick is heard to be even. This method can be used as an initial trial and will show how much adjustment needs to be made to the beat.

Many clocks have a special device installed to make the necessary adjustment, called a beat-setting device. Older clocks are less likely to have this facility unless they are of a higher quality. In some mantel, wall and regulator longcase clocks this takes the form of a fine screw which acts between the crutch and the upper part of the pendulum. Bringing mantel clocks into correct beat can be troublesome because it is necessary to turn the clock round to obtain access to the movement. If the surface on which the clock rests is not horizontal, making the adjustment when the clock is in a different position will not be effective. A good method is to test how level the surface is and then set the beat correctly with the clock on a similar surface to the chosen position. The clock may then be transferred to position, taking care to prevent the pendulum swinging about, thus altering the setting.

In some mantel clocks it is possible to attach the pendulum by hooking it on either forwards or backwards. Try both ways before attempting to adjust the beat, as there is likely to be some difference between the two and one may be in correct beat.

In most longcase clocks, and in some wall and mantel clocks, beat adjustment has to be achieved by carefully bending a thin rod (the crutch rod), which is specially intended for this purpose. If this is the intended method of adjustment there will be signs of prior bending. Special care has to be taken when doing this. The rod has to be bent *sideways only* and not *in* or *out*. Further, no damage must be caused to the clock, and it is clocks with this type of adjustment for which expert assistance is most likely to be needed. No load must be exerted on the escapement during the bending process. A series of minor adjustments should be made, bending to the correct side based on the initial testing, until the beat is correct.

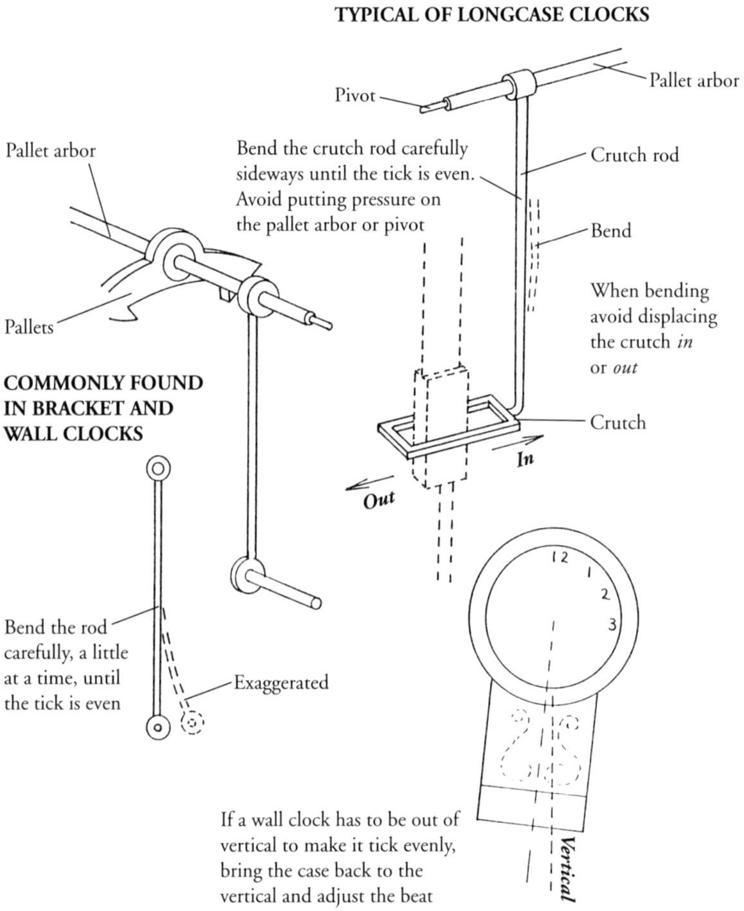

TYPICAL OF LONGCASE CLOCKS

Pivot

Pallet arbor

Pallet arbor

Crutch rod

Bend the crutch rod carefully sideways until the tick is even. Avoid putting pressure on the pallet arbor or pivot

Bend

When bending avoid displacing the crutch *in* or *out*

Pallets

COMMONLY FOUND IN BRACKET AND WALL CLOCKS

Crutch

In

Out

Bend the rod carefully, a little at a time, until the tick is even

Exaggerated

If a wall clock has to be out of vertical to make it tick evenly, bring the case back to the vertical and adjust the beat

Vertical

Fig. 4a   Means of adjusting the beat of a clock

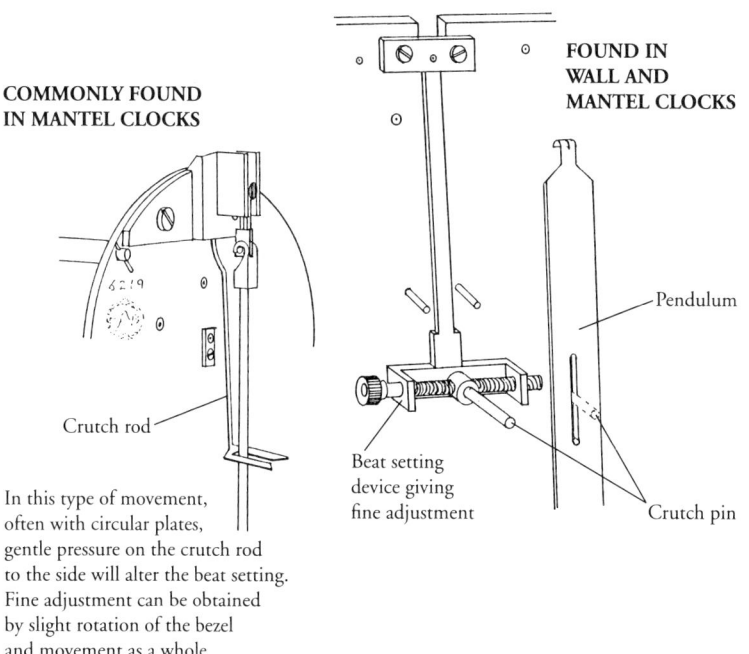

**COMMONLY FOUND
IN MANTEL CLOCKS**

**FOUND IN
WALL AND
MANTEL CLOCKS**

Pendulum

Crutch rod

Beat setting
device giving
fine adjustment

Crutch pin

In this type of movement,
often with circular plates,
gentle pressure on the crutch rod
to the side will alter the beat setting.
Fine adjustment can be obtained
by slight rotation of the bezel
and movement as a whole

Fig. 4b   Means of adjusting the beat of a clock

It will be found that the tick can be made perfectly even but it may not be maintained as the escape wheel of the clock rotates. The escape wheel normally rotates in sixty seconds and the beat will often vary slightly over this period. If the variations are very significant repairs to the clock may well be required to achieve a satisfactory going period.

Some mantel clocks have delicate escapements, particularly those which have the escapement visible in the upper part of the dial. In this type of clock, it is not possible or intended that the crutch rod should be bent. The clock is specially designed to allow the crutch rod to be slipped from side to side with only light friction. This is to protect the very delicate escapement from sudden jolts and to allow the beat to be easily reset. Moving this type of clock can easily cause slipping and the clock will go out of beat. Gentle pressure of the crutch rod, with the pendulum in position, to the appropriate side will change the beat. A few trials will quickly show how much adjustment needs to be made.

Winding a clock with a circular movement can sometimes cause the whole movement to rotate by a small amount if it is not securely fastened in the case. This will cause the clock to go out of beat and care has to be taken during winding to avoid this happening, but it can be used to benefit if the clock needs to be brought into beat. A small deliberate rotation of the movement within its housing in the appropriate direction will often be sufficient to bring the clock into correct beat. After making this adjustment the screws at the rear which retain the movement and the rear cover will need to be carefully tightened to avoid any subsequent movement.

Wall clocks were often fitted with a graduated scale level with the tip of the pendulum. This was installed for two reasons: to set the clock vertical and to help in setting the beat. The clock is vertical when the tip of pendulum is in the centre of the scale. This is achieved by moving the case. If the clock is not then in beat steps must be taken, as described above, to correct it. When correct the pendulum will swing an equal distance on either side of the centre of the scale. This can be a useful aid when the tick of a clock is very quiet or the owner is hard of hearing.

In 400-day clocks with rotating pendulums the clock should be set on a level and flat surface. The beat is very difficult to adjust in this type of clock and, it is a task for experts. A clock which has a balance wheel instead of a pendulum needs to be set in correct beat to tick evenly, and this also has to be done by a qualified horologist.

Regulator clocks have a small arc of swing of the pendulum and the beat needs to be set very accurately. A fine screw device is often fitted for this purpose. In this type of clock it is essential to ensure that the clock is wound before swinging the pendulum, otherwise damage may be caused to the escapement. This is also true of some other types of clock with delicate escapements.

Care needs to be taken to ensure that the weights are replaced in their correct position after reinstallation – the going weight on the going, the striking weight on the striking (for the hours) and, where fitted, the chiming weight (for the quarters) on the chiming. A thirty-hour longcase clock should normally have a weight of 7lb (3.2kg), an eight-day longcase clock will normally have two weights of 13lb (6kg). A heavier weight will usually be necessary to drive the chiming. Other sizes of weight are commonly found. When in good condition a clock will perform well with smaller weights. Heavier weights are sometimes used to encourage a sluggish clock to perform satisfactorily, but this is very bad practice, causing accelerated wear. The correct action is to maintain the clock properly.

When a weight-driven clock is set into position, an examination must be made to ensure that the fall of the weights is not affected by the front, back or sides of the case. A common problem in longcase clocks is the weights catching on the ledge at the bottom of the door during their descent or at the front of the case just below the seat-board. This may well be obvious when the clock is being wound. It is easy to check that the clock is not leaning forward and rectify as appropriate. It is also easy to see if the pendulum is rubbing on the back of the case and if the weights are coming into contact with the pendulum. Each of these failings can cause the clock to stop.

Where a clock has a separate wall bracket it is advisable to check that the bracket has more than adequate strength to support the clock, including during winding.

It is necessary to give the pendulum a gentle swing to start a clock, sufficient for it to start ticking. Ensure that the pendulum swings parallel with the back of the case, with the flat of the bob parallel with the backboard. A clock with a balance wheel which is in good order should automatically start when the clock is wound. Sometimes a *gentle* shake is necessary to cause the balance wheel to start to swing.

Winding

Opening the bezel or door for winding is normally straightforward. Longcase clocks can have a lock on the hood door which has to be released from inside the trunk door, but they are frequently incomplete. Mantel, bracket and wall clocks will have a catch, often key-operated in bracket clocks, to retain the door or bezel, allowing easy release. Cast bezels can sometimes be a problem to open. They can be very tightly fitting, requiring the insertion of a thin blade on the side opposite to the hinge if damage or disturbance to the clock is to be avoided.

Every care should be taken when inserting and removing the winding key. Damage is easily caused to the dial of a clock when winding, and such damage is often impossible to rectify. Only a key which fits snugly, with adequate engagement on the winding square, and operates without slipping should be used. A key which is loose or too large in diameter can bind on the dial, causing unsightly damage around the winding hole. If the key shows signs of splitting in the shaft or the handle it should be replaced before it breaks in operation. Keys are easily purchased for all types of clocks in a range of standard sizes. Many dials have winding holes fitted with brass or steel dial ferrules to protect the dial; if they are missing they should be replaced.

It is important to ensure that the clock is wound in one direction only, otherwise damage can be caused. Clocks wound from the front normally wind clockwise, whilst those wound from the back, such as carriage clocks, are normally wound anticlockwise. Arrows are sometimes provided to indicate the correct direction. It is very helpful when winding a clock to know how many turns of the key are needed for the full going period. If it is wound at the same time each week (or day) – and this is good practice – it will then require exactly the same number of turns of winding.

It is inadvisable to let the weights of a clock come to rest on the base of the case or the floor. This can lead to the rope or gut becoming dislodged from the pulleys if the weights fall over. The length of the rope or gut should still be sufficient to allow the clock to go for the full going period. When installing a clock it is best to arrange for the drop on the striking train weight to be a little greater than on the going train weight. This will ensure that if ever the clock is not wound the striking train will always be stopped in a safe position. In this way when the clock is rewound the striking will always be in the correct sequence.

Winding a clock fitted with chain, rope or gut lines can cause the weights to swing slightly. This can interfere with the pendulum and cause the clock to stop. It is good practice to keep an eye on the weights and pendulum when winding.

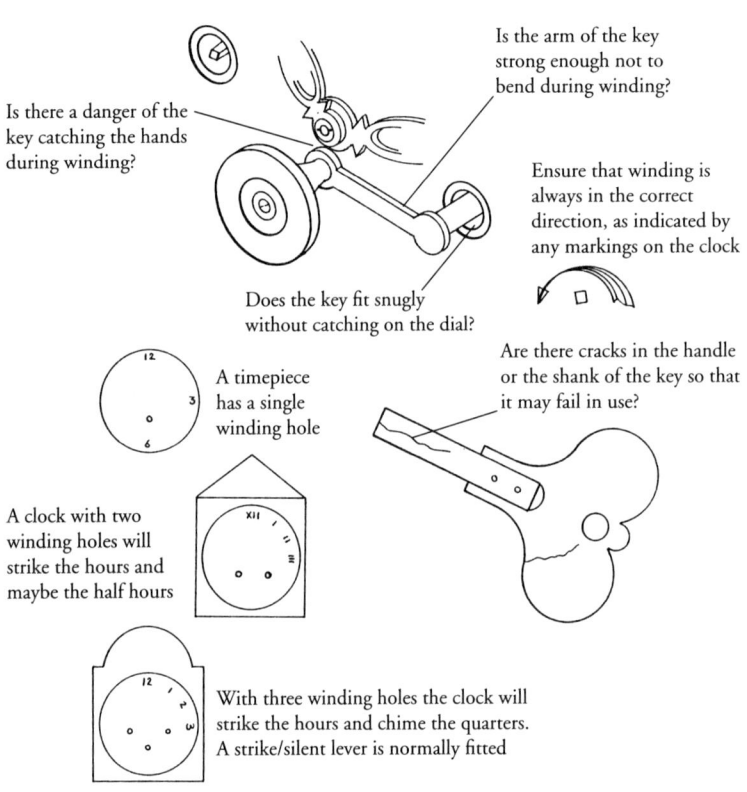

Is the arm of the key strong enough not to bend during winding?

Is there a danger of the key catching the hands during winding?

Ensure that winding is always in the correct direction, as indicated by any markings on the clock

Does the key fit snugly without catching on the dial?

Are there cracks in the handle or the shank of the key so that it may fail in use?

A timepiece has a single winding hole

A clock with two winding holes will strike the hours and maybe the half hours

With three winding holes the clock will strike the hours and chime the quarters. A strike/silent lever is normally fitted

Fig. 5a   Care in winding a clock

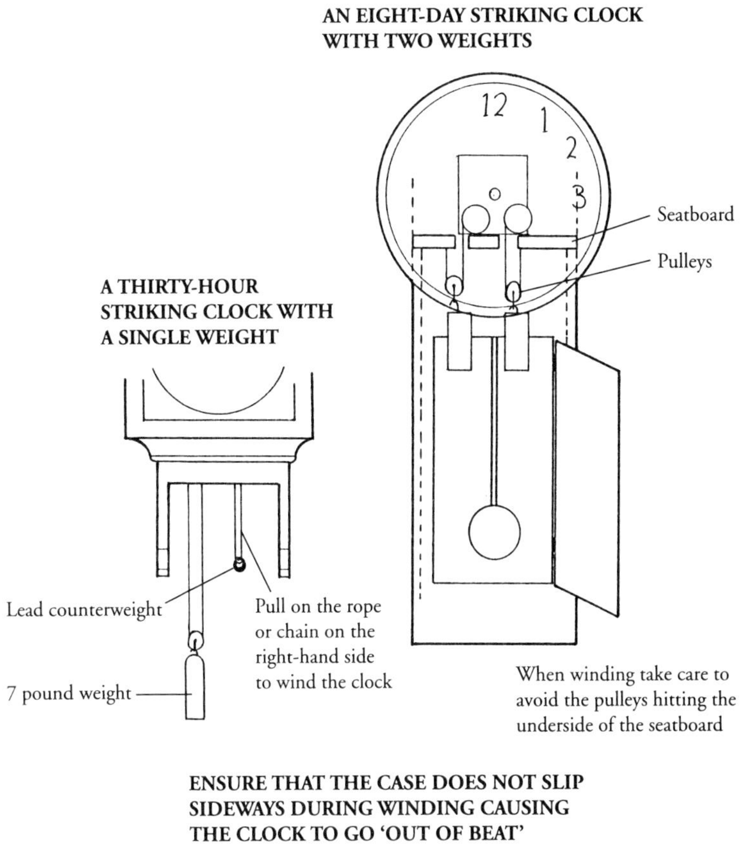

Care and Maintenance

AN EIGHT-DAY STRIKING CLOCK
WITH TWO WEIGHTS

A THIRTY-HOUR
STRIKING CLOCK WITH
A SINGLE WEIGHT

Seatboard

Pulleys

Lead counterweight

7 pound weight

Pull on the rope
or chain on the
right-hand side
to wind the clock

When winding take care to
avoid the pulleys hitting the
underside of the seatboard

ENSURE THAT THE CASE DOES NOT SLIP
SIDEWAYS DURING WINDING CAUSING
THE CLOCK TO GO 'OUT OF BEAT'

Fig. 5b   Care in winding a clock

Many spring driven clocks have no special measures to prevent overwinding and particular care therefore needs to be taken during winding. When a clock is fully wound any additional winding pressure is exerted on the end of the spring. Overwinding can thereby cause damage to the end of the spring risking serious damage to the delicate mechanism of the clock if it fails. It is for this reason that it is beneficial for the owner to have a regular pattern of winding and to know how many turns of the key it takes to fully wind the clock. During winding an increased pressure will be felt as the spring approaches the fully wound condition. Winding should then be stopped.

Some clocks, such as bracket clocks and some wall clocks, have a device (a fusee) to ensure that the driving power for the clock is constant throughout the whole of the going period. As a consequence, the effort required to wind this type of clock is constant. There is therefore no warning that the clock is close to being fully wound. This type of clock is fitted with a device which prevents further winding when it is fully wound, but care still needs to be taken to avoid overstressing this protective device.

When winding a clock with weights, if the pulley hits the underside of the seatboard it can be upset, allowing the rope, gut or chain weight to come off and permitting the weight to drop a short distance. This will stop the clock or the chiming, whichever is being wound, and will cause damage to the rope or gut. It is advisable always to wind the weights up to the same position, allowing a few inches of clearance between the pulley and the seatboard. Suitably placed chalk marks on the weights are sometimes used as a way of ensuring consistent winding. It is desirable to inspect the rope, gut or chain from time to time. Any slight damage should be examined to determine whether it is liable to fail in service. When gut becomes hard or has cuts or loose strands it should be replaced. This is most important on spring-driven clocks, where a gut failure can lead to considerable damage to the movement. With a weight-driven clock the potential damage is to the floor or the base of the clock if the weight falls. When winding the gut should wrap neatly on the barrels without overlap. If a rope is installed for winding, loose strands and fluff can enter the mechanism and a replacement should be fitted when signs of wear become apparent. Where a chain is installed the links should all be closed, with no damage. Replacement rope, chain and gut is easily obtainable.

It may be necessary to support a heavy metal bezel during winding if there is either a risk of damage to the hinge or a chance that the glass may fall out.

When it is not possible to wind the clock before the end of its going period a decision has to be made whether to stop it or to let it run until it stops. In weight-driven pendulum clocks it is often easiest to stop the clock and restart it when required. This avoids the possible problem of the striking becoming out of the correct sequence. In spring-driven clocks with pendulums or in clocks which do not have pendulums it is wisest to let the clock run without interference.

Setting to Time

The minute hand of a clock can be turned forwards without risk of damage provided the going and striking are both partly wound. It is safer to hold the minute hand near to its centre to avoid any risk of bending or breakage. No attempt should ever be made to move the hour hand on its own. It is necessary to let the clock chime fully at each hour and half hour (if appropriate) and particularly so at twelve o'clock. Many clocks have a mechanism to prevent a problem arising if striking is not completed, but it is not always effective. The hands on a striking or chiming clock must not be turned backwards (unless you have expert knowledge of the clock mechanism). However, if the clock is a few minutes fast it is always safe to turn the hands back from the 'five minutes past the hour' position to the hour, but *not* past the hour. This is because the striking is primed just before the hour and damage to the mechanism may result. If the clock is significantly fast it is advisable to stop it temporarily until the correct time is reached.

Care should be taken when moving the hands of a clock that marks are not made on the dial. It is better to avoid any contact with the dial, as in time this will lead to loss of the numerals and markings, and may cause the onset of corrosion.

After the hands have been moved, make sure they are not bent out of place so that they risk touching each other, the glass or the dial. The minute hand or seconds hand may catch on the hour hand. These are common causes of a clock stopping.

When the minute hand is at the hour it is advisable to check that the hour hand correctly points to the centre of the hour numeral. In some clocks the hour hand can slip independently, allowing careful adjustment, but, in most it is permanently fixed. Any correction therefore needs to be carried out by someone with the necessary expertise.

If a hand is fitted to record the date, it is usually set to change over at, or just after, midnight. It may not be possible to reset if the date is in the process of moving.

When changing over from winter to summer time it is necessary to move the clock forward by one hour. This can be done by slowly moving the hands forward, allowing any chiming to be fully completed. When moving from summer to winter time it is necessary to set the clock back by one hour. The safest and most convenient way of achieving this is often to stop the clock and restart it after an hour has passed.

The hands are driven by a simple friction device so that the clock continues to go whilst the hands are moved. If the friction is inadequate the clock will not drive the hands forwards. If it is too tight, hand setting will be difficult and the owner may cause damage to the hands if excessive force is used when setting to time. Specialist advice will be needed to rectify this problem.

Timekeeping

All clocks are provided with the means to adjust the going rate. In the case of a clock with a pendulum this is achieved by altering the effective length of the pendulum, usually by moving a nut at the top or bottom of the pendulum bob. Movement of the nut to make the bob lower will make the clock lose time and vice versa. Start by making small adjustments, such as half a turn of the nut, and monitor the influence on the rate over a day or two. Larger or smaller adjustments can then be made as required. Long pendulums give more accurate timekeeping. Fine adjustment is easier; the longer the pendulum the greater is the change in length required to achieve a given change in the rate of the clock. Short pendulums are sensitive to small changes in length. In some larger clocks the effective length of the pendulum is altered by adding or removing small weights to a pan attached near to the top of the pendulum rod.

Mantel and bracket clocks are sometimes provided with an additional means of fine adjustment. This is operated by a small square or hand at the top of the dial, linked to a device to raise or lower the suspension spring for the pendulum. This avoids the need to turn the clock round to see into the back – a very useful feature. Turning the square with a small watch key has the desired effect. A ratchet on the device allows adjustment in small discrete increments. It is best to experiment to be certain which direction makes the clock gain or lose. Usually, turning the square clockwise makes the clock gain.

To adjust the rate rotate the
nut at the base of the pendulum;
lowering the bob slows the clock

In longcase clocks one full turn
of the nut alters the rate by
about $^1/2$ minute per day

**COMMON IN LONGCASE,
WALL AND MANTEL CLOCKS**

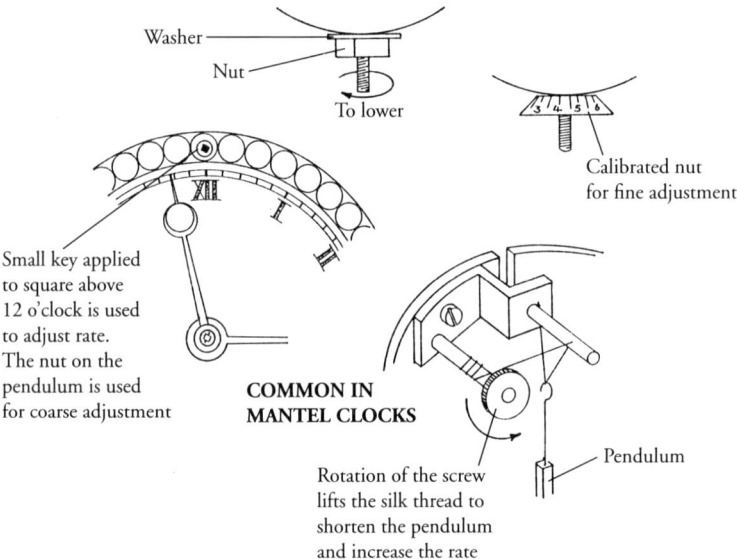

Washer

Nut

To lower

Calibrated nut
for fine adjustment

Small key applied
to square above
12 o'clock is used
to adjust rate.
The nut on the
pendulum is used
for coarse adjustment

**COMMON IN
MANTEL CLOCKS**

Rotation of the screw
lifts the silk thread to
shorten the pendulum
and increase the rate

Pendulum

Fig. 6a   Means of adjusting the rate of a clock

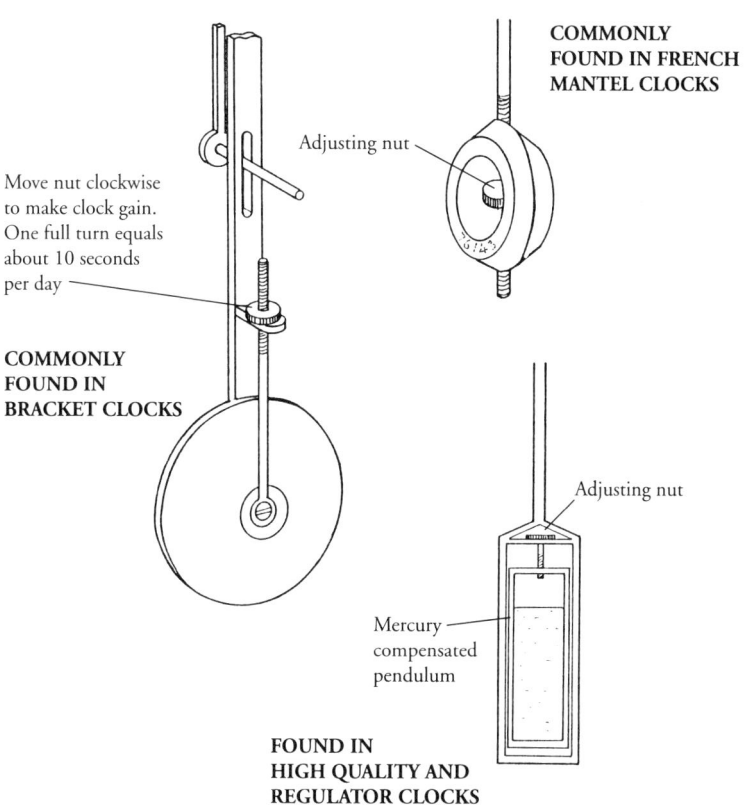

COMMONLY
FOUND IN FRENCH
MANTEL CLOCKS

Adjusting nut

Move nut clockwise
to make clock gain.
One full turn equals
about 10 seconds
per day

COMMONLY
FOUND IN
BRACKET CLOCKS

Adjusting nut

Mercury
compensated
pendulum

FOUND IN
HIGH QUALITY AND
REGULATOR CLOCKS

Fig. 6b   Means of adjusting the rate of a clock

Clocks which have a balance wheel require adjustment in a different way. It is necessary to alter the effective length of the balance spring to alter the period of swing of the balance. A lever or ring on the balance assembly is provided for this purpose, often with a graduated scale. Only small adjustments are necessary. Particular care is needed due to the fragile nature of the balance assembly. If it is not possible to achieve a satisfactory rate, expert assistance will be essential.

The rate of a clock with a balance wheel is usually insensitive to changes in room temperature, but the rate of a pendulum clock is affected to some extent. In warmer conditions expansion of the pendulum rod causes the clock to lose, and the shorter the pendulum the more the accuracy changes with temperature. A small adjustment may be necessary between summer and winter. Some clocks have temperature-compensated pendulums and these should require little or no adjustment.

Striking the Hours and Chiming the Quarters

Many clocks are provided with a lever to turn the chiming/striking on or off. When such a lever is not provided, the striking and chiming should be wound up and in operation whenever the clock is going. In many clocks, if the chime is not wound up and operating, the clock can stop.

When the clock is in the process of striking, or in the period just before striking, the 'strike/silent' lever must *not* be operated. It is advisable to operate this lever when the clock has just struck. When the 'strike/silent' lever is subsequently moved back to the 'strike' position, the clock may initially strike incorrectly. Some clocks have an inbuilt system which returns the striking to the correct sequence at the next hour, but most do not have this facility and steps are then necessary to correct it manually. Fortunately clocks are designed to make this easy to accomplish.

The method to bring the striking back into sequence with the hands depends upon the type of clock.

• **Thirty-hour clocks.** This category includes some longcase, lantern, mantel, wall and hooded clocks. An antique clock which runs for thirty hours will normally employ a single weight, using the 'Huygen's endless rope' system which drives both going and striking. Some later clocks may be spring-driven. To economize on use of the driving power these clocks are able to strike the hour once only in the correct sequence, but cannot repeat the hour. The sequence is controlled through a locking plate which rotates by the correct number of notches. It is easily recognized by the uneven spacing of the notches or pins around its perimeter. One notch or pin equals one blow on the bell or gong. A short hooked lever, a detent, hangs above the plate, and after striking the correct number of blows it drops into one of the notches, locking the plate. In effect the plate is a program, which the clock reads to indicate how many hours are to be struck. Clocks which chime the quarters can use a similar type of programmed plate.

The striking can become out of sequence in two different ways; the hour fails to strike for some reason or the hour is induced to strike prematurely. To rectify the problem it is necessary to induce the clock to strike the next hour or hours until the correct position in the sequence is reached.

A lever is usually installed on the movement which, when lifted, causes the striking train to be released. This is often the same lever which is lifted on the hour as the minute hand makes its hourly circuit. An extension piece is sometimes fitted to the lever allowing it to be pressed downwards. When viewed from the rear it is often visible at the right-hand side of the movement. It has to be pressed down firmly but carefully and then released. When the clock has fully completed its striking the process is repeated until the correct hour is reached. When the lever is not provided it is necessary to lift and release the detent instead. This releases the locking plate in exactly the same way.

**COMMONLY FOUND IN
STRIKING MANTEL CLOCKS**

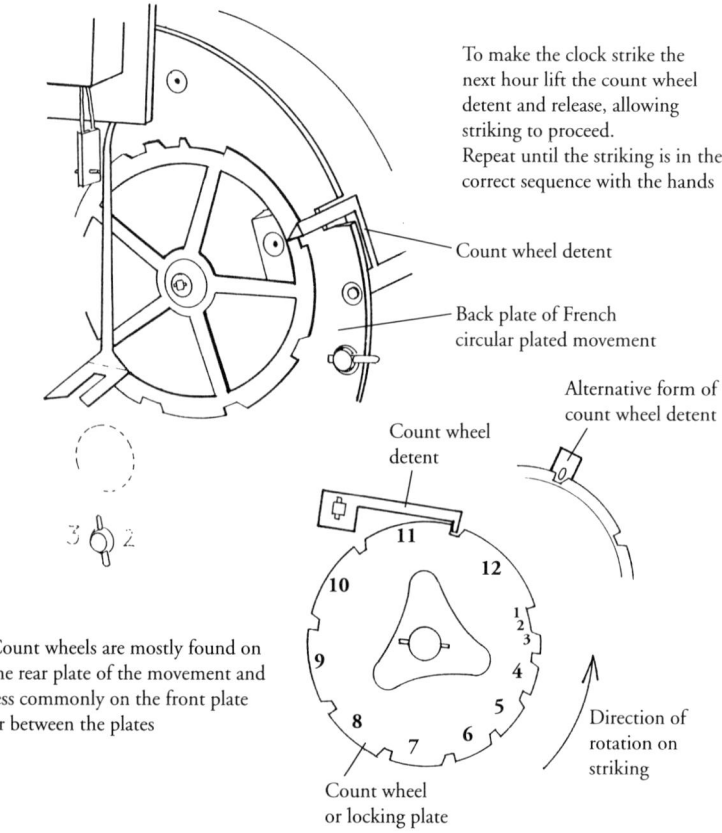

To make the clock strike the next hour lift the count wheel detent and release, allowing striking to proceed.
Repeat until the striking is in the correct sequence with the hands

Count wheel detent

Back plate of French circular plated movement

Alternative form of count wheel detent

Count wheel detent

Count wheels are mostly found on the rear plate of the movement and less commonly on the front plate or between the plates

Direction of rotation on striking

Count wheel or locking plate

Fig. 7a   Count wheel striking – how to bring a clock back to the correct striking sequence

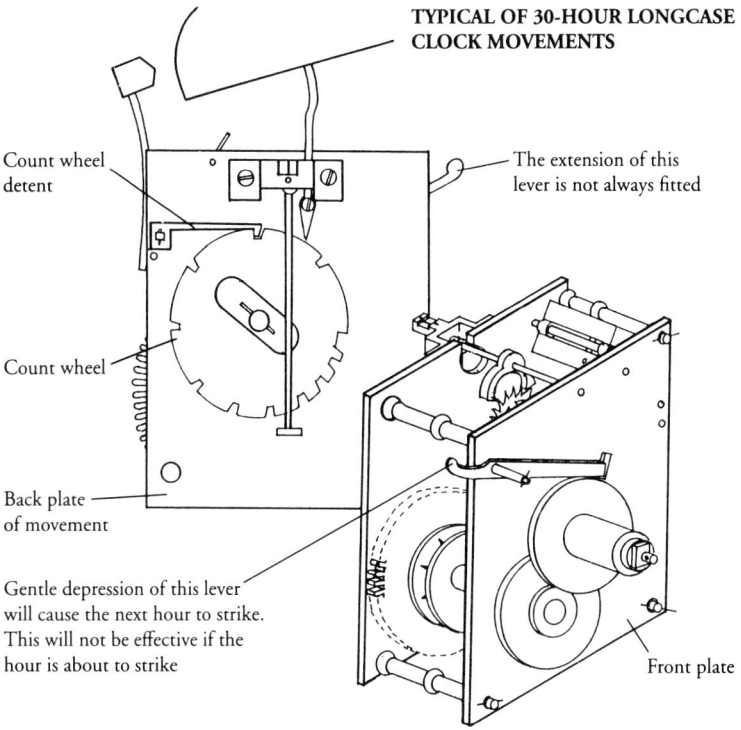

TYPICAL OF 30-HOUR LONGCASE CLOCK MOVEMENTS

Count wheel detent

The extension of this lever is not always fitted

Count wheel

Back plate of movement

Gentle depression of this lever will cause the next hour to strike. This will not be effective if the hour is about to strike

Front plate

Fig. 7b Count wheel striking – how to bring a clock back to the correct striking sequence

• **Eight-day clocks.** Eight-day clocks have a larger stored energy to provide for the extended going period. In fact, they usually have an excess of stored energy, allowing them to strike the hour more than once if required. However, not all eight-day clocks have this feature. Those which do, have a striking system called rack striking. This also uses a programmed plate, but it operates in a different way to the locking plate for thirty-hour clocks. The 'program' is in the form of a snail-shaped disc rigidly attached to the tube supporting the hour hand, which contains the information to tell the clock how many hours to strike when the hour hand is pointing at any hour. The rack is the reading device. On the hour it is activated and reads the disc so that the correct number of blows are struck. The same type of system is used in some high-quality clocks which chime on the quarters. This type of system cannot get out of sequence. It can only be at fault if the hands are put on the clock in the wrong position for the programmed disc. (If a clock strikes erratically or too many blows the problem normally lies with the reading device. Repair is then required.) Some eight-day clocks, and some mantel and wall clocks, have a striking system similar to thirty-hour clocks and therefore can become out of the correct sequence.

• **Clocks chiming the quarters.** When a clock which chimes the quarters is allowed to stop the quarter chiming as well as the striking can become out of sequence with the hands. Some mantel clocks have a system which automatically brings the clock back into the correct sequence at the next hour.

• **Repeating clocks.** A pull-repeat cord or lever is normally provided to allow repeating of the last hour. In many types of clock normal striking will be prevented in the few minutes before the hour. With some mantel and carriage clocks the repeat can be operated right up until the hour. In both types the hour struck will be the previous hour.

If a clock continues to go but regularly fails to maintain the correct striking or chiming sequence, cleaning and lubrication or other maintenance is required.

If the clock is intended to strike and/or chime but does not have a 'strike/silent' lever to silence it, it is possible to have the striking inhibited. This usually requires a minor change to the movement by someone with expert knowledge. It can subsequently be reversed without harm to the clock. In clocks provided with the endless rope systems, such as thirty-hour longcase clocks with a single weight, this modification will extend the going period between windings to about three days.

Clocks which have the facility to chime different tunes or chime on four or eight bells or gongs will have a lever or hand to make the selection. This lever must *not* be operated whilst chiming is in progress. It is best to operate it immediately after chiming has been completed.

There is usually some scope to alter the sound of the bells or gongs where the sound is unattractive or a nuisance. This is usually related to the way in which the hammer contacts the bell or gong, and how securely the bell or gong is attached. It is a matter which may require expert assistance.

Dust

Clocks are usually provided with a case or cover, whose primary purpose is to exclude moisture and dust. Means of access are then provided in the form of hinged covers or doors. Silk or other material is used inside wooden or brass frets to allow the striking to be heard whilst still preventing the ingress of dust. To ensure satisfactory long-term running, all means of excluding dust should be kept in place at all times. Where there are gaps or openings in a case which will admit dust they can be minimized with adhesive paper or strips of packing. Any material employed should be capable of being removed without damage to the case.

Cleaning

The Case

An occasional rub with a good wax polish and a regular dusting should be sufficient to keep a wooden case looking clean and bright. Soft wax polishes are very good at loosening ingrained dirt and grease. It is advisable to avoid any abrasives or treatment which will harm the original patina of the case. Ormolu or gilt items should be carefully dusted to avoid any risk to the thin coating.

Regular polishing of brass with ammonia-based agents can lead to problems. First, the fine abrasive is hard to remove and leaves unsightly deposits in corners and crevices. Secondly, the ammonia causes weakening of the brass, particularly in porous cast brass or items formed from thin sheet. Brass items should therefore be thoroughly cleaned and then lacquered to avoid regular polishing. This is an easy task for the professional restorer or the enthusiastic owner.

The Glass

A sparkling glass greatly enhances the look of a clock. It can be cleaned with a domestic cleaning agent or a personally preferred method. Care must be exercised in two matters. First, the glass in some frames and bezels is often loose, so care needs to be taken to support the glass from behind when cleaning. Domes and domed circular glasses are particularly vulnerable and are also costly to replace. Secondly, avoid dripping water or cleaning fluids onto the case, particularly when it is of wood.

The glass panels in carriage-type clocks can easily be cleaned externally, but care needs to be taken to avoid any liquids entering the case. Cleaning the internal surfaces requires removal of the movement, which is a tricky operation. It is therefore best to leave the internal surfaces of the panels until the next repair or maintenance. The corners of a glass often retain dirt and sometimes traces of varnish. These can be carefully scraped out with the edge of a metal object prior to cleaning, and a much-enhanced appearance is achieved. Painted glass panels should be treated with the greatest care; no action should be taken on the painted side.

The Dial

As the final stage of being restored, brass and silvered dials should be lacquered to avoid tarnishing. Occasional and very light dusting is then all that is required. Regular polishing of unlacquered brass dials should be avoided since this causes wear to the dial as well as unsightly deposits around it.

Painted, enamelled or plastic dials should be lightly dusted only if it is essential, taking care not to use any cloths impregnated with wax or other polish. The use of spray polishes in the vicinity of the dial is inappropriate. The black numerals and markings will often be in poor condition but if they are still serviceable it is best to retain them in their original form. Cleaning will not extend their life, since they are often marked in a water-soluble ink or other fragile medium. Any oil or grease on the dial should be carefully absorbed to prevent it spreading over the surface. Enamel can be safely cleaned with clean water to soften the grime provided the owner is certain it is enamel and not paint or other material.

The Movement

If a clock has congealed oil around the pivot holes, is completely dry of oil, or even worse, has oil and dirt around the pinions, professional cleaning is required. The application of fresh oil to a dirty movement accelerates wear. Signs of rust on the steel components are another indication that cleaning is required. Cleaning should be carried out by someone with appropriate knowledge and followed by correct lubrication.

Storage

When not in use a clock should preferably be stored out of direct sunlight, in a dry, warm and adequately ventilated environment. Care should be taken to prevent it becoming too hot or dry so as to reduce the risk of shrinkage of wooden cases and to minimize the deterioration of the lubrication.

A clock must *not* be stored in a plastic bag or tightly sealed container, as this may cause condensation on the steel components. In due course rust will occur, causing deterioration and serious long-term damage. It is also advisable to take steps to prevent dust entering the clock during prolonged storage.

Spring-driven clocks should be stored with the springs wound down to prevent them deteriorating.

Insurance and Security

With the growing value of antique items and the increased risk of theft, it is important that the owner of a clock should be aware of its current value for insurance purposes, and ensure that the level of cover is adequate. It may already be included in a household insurance or 'all risks' policy, but where an owner has a collection of clocks the insurance company will usually require a list of all the items in the collection, with a professional estimate of their value. Without this there may be a limit set on the amount of any claim.

Once a professional valuation has been obtained, annual increases in line with inflation can be made to keep the value up to date. Top auction houses now recommend that valuations are obtained at three or four-yearly intervals for fine art items, to ensure that premiums are adequate but not excessive. Discounts on insurance premiums are sometimes available when up-to-date valuations are available and accompanied by a photographic record.

Competition for business has made premiums more competitive and also more flexible. As a means of optimizing the type of cover and the cost, it is now possible to arrange special cover to suit particular needs. The owner may wish to insure large items such as furniture against fire only, whereas full cover may be more appropriate for small items such as clocks or watches. It is necessary to agree with the insurer the basis of the cover: whether it is the current worth on the open or auction market or the cost of complete replacement at current retail replacement values.

Full details, including photographs, can be invaluable in the case of a claim. It is useful to include a ruler in photographs so that the actual size of the item can be judged. Full details of a clock, with records of any distinguishing marks or features, can be important when proving ownership or aiding recovery.

Finally, when an item is sent for repair, some form of receipt should be provided by the repairer.

Repair and Maintenance

A clock which is kept running continuously, and is expected to do so for many years to come, should be maintained on a regular basis.

This will generally involve cleaning, lubrication and the rectification of accumulated wear. Advice on restoration and restorers is included in detail in Chapter 3.

Lubrication

All clocks should be kept properly lubricated. Some parts need no lubrication, some require heavy oils and some fine oils. This is a specialized matter which, if it is not done correctly and at the correct frequency, can affect the satisfactory operation, timekeeping, safety and long-term life of the clock. Ideally, lubrication should be carried out every five years, depending on the type of clock, its environment, and the effectiveness of its case at keeping it dust-free. Once the lubrication ceases to be effective, the wear on the clock will be accelerated. It may often be necessary to clean the movement before lubrication can be correctly applied. As the friction in the going train increases, due to the deterioration in the effectiveness of the lubrication, the angle of swing of the pendulum will be reduced slightly. This will lead to a gain in the rate of the clock. Evidence of such a change is a sign that lubrication or other maintenance is required.

Repair

It is useful to a repairer if the owner has kept a record of how and when a problem occurred. When there is a malfunction, it is best to resolve the matter before it leads to additional difficulties. For example, a problem with the striking of a clock may lead to problems with the going. If the owner wishes to attempt to carry out any repairs it is advisable to study one of the available books on this topic before going ahead. These will provide advice on tools and materials required and the skills which will be necessary.

2 Buying and Selling

Buying

Having decided upon the size, shape, style and type of clock which will meet the potential buyer's needs there are three other matters which are important: how to assess the cost or value of a clock, where the best place is to buy, and how to assess if the clock is in going order. There is guidance on assessing the condition of a clock in Chapter 6; this section considers the cost and where to buy.

The Cost or Value

The cost and value of a clock very much depends on the market situation and the perceptions of the seller and buyer. Full consideration of the value of an item is given at the end of this section. The cost is easier to assess, being composed of two components. These are the purchase cost and the cost of bringing the clock into a satisfactory going condition. The purchase cost is easy to resolve but the second is a different matter. The costs of carrying out any maintenance, repairs or restoration will be strongly dependent on the condition of the clock and the extent to which it is complete and original. This is a high-risk area for a potential buyer, and it is highly desirable to be able to quantify it at the time of purchase, as it could vary between nothing and a large sum.

When purchasing an antique or collectable clock it must always be remembered that it is second hand. It may operate perfectly, or it may be completely inoperable. One thing is certain – if it has been in use for many generations then it will have seen many different repairers over its life. Considerable work on it may have been done. If it is to be used as a timekeeper as well as an attractive domestic item, a proper assessment of the effort and cost required to bring it back into good going order is essential. The would-be purchaser could take a gamble but would be better to take expert advice. Fortunately, antique clocks were generally made to last for a prolonged period. That they have done so is testimony to their design and construction.

It is beneficial if the intending purchaser is in a position to make a general overall assessment. This will ensure that items which are not worth serious consideration are immediately rejected. Having made this initial selection, if the risk remains significant, expert advice can then be sought. This will avoid the unnecessary effort and expense of assessing every clock which takes the potential owner's fancy.

A general assessment will need to include the condition and completeness of the clock: the current state of repair, the operation of all its functions, and any missing components. In addition, it must embrace as thorough an examination as is practicable of whether the main parts are original or a marriage of items from more than one clock. Originality is of particular concern since it will be important in relation to both the value of the clock and the cost of restoring it. Guidance on this is given in Chapter 4.

The buyer has three options available when considering where to buy a clock: the private sale, the auction room or the professional dealer. Which to choose is a matter of personal choice, but each has different risks and conditions attached.

Private Sales

Some of the best items never appear in the market place and are sold privately. This is often the best way to obtain a quality piece at a fair price. Unfortunately, items do not appear very frequently among one's contacts and acquaintances. Private advertisements are another potential source, although the private sale may be a professional sale in disguise.

In a private purchase the buyer has no legal recourse if the item turns out not to be as expected, although if the item is subsequently found to have been stolen, the buyer is likely to have a good case to recover his money, provided that the seller can still be traced! Some proof of purchase is desirable, including details of the seller.

The price paid will be by mutual agreement with the seller – let the buyer beware!

Professional Dealers

Many more clocks will be on offer from professional dealers. An important benefit is that the terms of purchase are much better defined. A dealer who is a member of a recognized professional body has an obligation to offer a correct description of the item. It is essential to obtain a detailed invoice, spelling out the details of the age, type and condition of the clock. If the clock proves to be a bad performer, a fake, or not as described, then the owner has considerable recourse. If a guarantee of satisfactory performance is offered, this should be detailed on the invoice, in case of future dispute. When buying from other traders or dealers the situation is little different from private purchases. Your only possible safeguard is the reputation of the dealer in the eyes of the public.

In general, dealers in horological items will offer items which are restored and in good operating condition. This is because they can sell them for a higher price. Such an item will usually be offered at a fixed price and with a guarantee. An unrestored item bought from a dealer carries no type of guarantee and therefore involves a large risk for the unwary buyer. The repair or restoration costs may be small, but they could be very large. Further, if the clock is a marriage of items, some of which are not original, it may be impossible to restore the clock into a satisfactory condition. The sale of an unrestored item by a dealer specializing in restored clocks should always be viewed with an element of caution.

Antiques and collectable fairs offer the opportunity to view a large number of clocks of different types with the minimum of effort and travelling. At the top of the range are prestige fairs selling high-quality items, usually fully restored, at full commercial prices. At the other end are fairs and markets, some of which specialize in clocks and watches, where a majority of the items are unrestored. The potential buyer should always remember that there is always scope for the negotiation of prices at these specialist fairs. Dealers will often be happy to sell an item at a sensible profit rather than allow their capital to languish for long periods. They usually have items which they would very much like to buy if only they had a little more capital to invest! Another benefit of the fairs and markets is that a full range of original and reproduction spares for use in restoration and repair are usually on sale. This will interest both the restorer and the dealer but has a great deal to offer the owner as well.

Auctions

Auction houses will offer many horological items in their general sales and will also have specialized sales of rare and high-quality items. At auction, buyers have the benefit of the auction house providing a description of the item in their catalogue. They will be able to read the description of the particular item and compare it with their own observations. Of particular importance is the manner of description used and the intending bidder needs to be familiar with the subtle differences in meaning. The terminology will be explained in detail in the catalogue, and if the description proves to be erroneous or outside the stated terminology, the purchaser has considerable recourse. In addition the auctioneers and their staff will be happy to offer advice and assistance when requested.

Whilst the description will usually be accurate in terms of the age and style of an item, however, it will not necessarily give other important information. The auctioneers may not have the time or the expertise to judge the originality of its component parts, or its completeness, and intending bidders need to make their own assessment of these aspects.

The benefits that the auction house offer are balanced by the charges which are made. A commission of 10–15 per cent on the hammer price (plus in Europe VAT on the commission) is fairly common. Before bidding it is useful to calculate the total price to be paid, including commission for your intended bids. This ensures a firm grip on reality. It is necessary to bear in mind that the auction house is working to sell an item on behalf of a client, and it may well present the item in its best light to obtain the best sale price. Note, too, that as with purchases from dealers, unrestored items bought at auction carry no guarantee.

Selling

The same three options for buying are also open for selling.

Private Sales

For the seller, the shortcomings of a private sale are fairly clear. A customer has to be found. A good idea of the market value for the item is required. It will be necessary to negotiate with the prospective purchaser over the price. Unknown people may have to be allowed access to the home. A cheque or a pile of unknown banknotes will have to be accepted. It can be quick and painless, but it can also be very stressful. To make matters worse the new owner may return to insist that the clock is not what he or she was led to believe, or demand a refund of money.

Advertising of some form will be necessary to alert the world to the item. Viewing will have to be arranged at mutual convenience. One useful technique when selling an item, where practicable, is to put it outside the premises, e.g. in the garage, before the customer arrives. In this way many potential security hazards can be avoided and if a deal is done, shipping out is easy, with no risk of damage to the paintwork or other possessions. It is always easiest to sell one item at a time. This avoids the purchaser trying to obtain a reduction for quantity. It is beneficial for both parties for the purchaser to see all aspects of the clock, including any shortcomings. The opportunity should be taken to demonstrate the operation of all aspects and any special features or peculiarities. This is the only way of reducing the risk of the buyer feeling dissatisfied later.

Professional Dealers

There are many professional dealers, both with and without high-street premises, who make their livelihood buying and selling clocks, watches and related items. Their business inevitably succeeds by maximizing the difference between the buying and selling prices. Dealers who belong to a professional body may be bound by certain rules of good conduct, but whilst this may bind them to offer correct advice regarding age, type, provenance and originality, it does not preclude them from offering low prices for items. It is in the best interests of the clock owner when considering a sale of a valuable item to seek one or more offers, preferably from different types of purchasers. Even so, each of these may be well below its true market value and it is worth considering other routes to a sale.

A dealer or other potential purchaser may ask about the legal title of the item and may ask for some form of identity or proof of ownership. If these are easy to provide, it will give the purchaser the necessary level of confidence.

One basic piece of advice is *never* to accept the first offer made by any potential buyer, even if it appears to be generous. Offers are usually 'dressed up' in some sort of persuasive patter – 'It is the most that can be offered' or 'It is in very poor condition for an item of this type', or 'The market for such items is not good at the moment'. The wise owner will ignore these completely. A first offer is usually a 'sighting shot' to judge the reaction of the seller. It usually signifies a serious interest in buying and the intention of offering a higher figure. It is best to parry an initial offer by asking 'How long is the offer valid for?' Always resist as much as possible the temptation to make a quick decision, unless you are desperate to sell. With a little patience the price could easily be increased by 50 per cent or more. Indicate that another offer is to be obtained or, even better, awaited from elsewhere, and assess the reaction. If a better figure is offered it is still wise to indicate that a decision will be made at a later time. If the bidder is keen then interest will be shown in another way. A note of the phone number and a promise to make contact in a few days is a useful step. Only when the owner feels reasonably confident that the best offer has been obtained should the decision to sell be made.

A professional dealer will often prefer to buy items which are either in an unrestored condition or in good going order. In the first case the dealer is offered the opportunity to improve the item by appropriate repair and restoration and thereby sell it at a higher profit margin. If the dealer carries out the work this allows a sale of their own labour at a premium rate. Clocks in good order are attractive because they require almost no effort to put them imme-diately up for sale. It is the in-between items which can offer the most problems; partly or poorly done work may need to be corrected at extra cost. If the external finish of the case has been removed or 'improved' then it can never be replaced and this will deter both trade and private buyers.

Auctions

The third option is to approach a reputable auction house. Sale rooms usually hold regular sessions when potential sellers can seek free advice. This will usually include an opinion on the saleability of the item and an estimated sale price. There is no obligation to sell but there may be an invitation to do so. The service works on the basis that the goodwill generated will benefit both parties in the long run. The estimate provided by the resident valuer will be towards the lower end of the likely range of selling price. This has advantages for both parties. It avoids any disappointment if a higher price is not achieved on sale and makes the seller very happy with the outcome and with the auction house if it is.

At auction the seller will incur the standard charges of the auction house. These cover the advertising, insurance and sale of the item in the wide public forum. These charges are currently of the order of 10 per cent of the selling price, (plus in Europe VAT on the commission). The conditions set by auctioneers regarding sales and failure to sell are usually described in full inside their cata-logues. It is worth studying these in detail. Failure to sell at an agreed reserve may lead to a further opportunity at another sale, at a lower reserve, at no extra cost.

One of the primary benefits of selling at auction is that it provides a current market price when the owner is out of touch with the value of an item. Many times people have been very grate-

An example of a bad repair to veneer on the lower cross-banded edge of the door and the herringbone edging. The new surfaces are not flat, the matching of the colour and grain is poor and the grain is open – all helping to accentuate the repair

Mid eighteenth-century oak case with pagoda-style top. The dial fits neatly inside the dial frame with the spandrels fully in view suggesting, but in no way guaranteeing, that the dial and movement are original to the case. The quarter round beading on the internal edge of the mask is completely original

A shallow recess is cut into the front of the trunk of a good quality late eighteenth-century clock case to retain the dial, showing that the dial is the correct size and helping to confirm its authenticity

An example of a mid nineteenth-century English longcase dial with a well painted 'false' moon dial in the unusually large arch. Clearly, the moon is intended to impress but not to operate

ful they used an auction house instead of an apparently generous offer from a neighbour or friend to 'take the item off their hands'. A further benefit of using an auction house is that a sale is achieved with very little effort from the owner, although the whole process may take a few months before a cheque is received.

Experience shows that the price realized at a sale is partly related to the type of sale. If the clock is the only antique item in a general agricultural sale, for example, there may not be sufficient interest in clocks to achieve the best price. The owner may be lucky, but it is best to find a good antiques sale where there will be a number of other clocks and watches of similar or better quality. If the item is particularly rare or of a high quality it is worth considering a sale specializing in horological items only. This can be discussed with the auctioneers. It is often advisable to be prepared to wait for a suitable sale in their local or national programme.

Auction houses usually advise sellers not to have an item restored prior to sale. Their experience is that unrestored items sell as well, if not better, than restored items.

A final word of advice. Are there any supporting documents giving details of the purchase or history of the clock, or its previous owners? These can add to the potential sale value, particularly if there is an unusual history or an owner was a well-known person. In such instances, selling in an auction room is likely to be the most advantageous route. The auctioneers will be happy to use maximum publicity for mutual benefit.

The Financial Value of a Clock

Whether buying or selling a clock a knowledge of the current market value is very useful if a sensible price is to be paid or obtained. There are various ways of obtaining advice and guidance.

Information on the value of a clock can be obtained from the various published price guides. The prices are normally based on those realized at recent auctions. They are very useful, particularly when comparing the cost of different types of clocks, but cannot give a realistic measure of the quality, condition or originality of the items. In addition, the prices are always based on past sales. Market conditions may well have changed significantly since the time of preparation and publication of the guide.

When buying or selling at auction the resident valuer will give advice regarding a reserve price and offer an opinion on the likely hammer price. Because the item is being sold in a public forum, provided that the sale has been properly advertised, a reasonable market price will be achieved. Much has been written about the illegal practice of dealers' rings not bidding against each other in order to buy at a low price. They still occur, but fortunately, only a small number of dealers will belong to each ring and therefore at a good auction room its effect will be largely nullified by other dealers and private buyers. It has been the practice of auction houses to encourage private buyers into their sale rooms. This has benefits for all honest parties. It protects the seller, it nullifies the rings and it maintains a stable market. For dealers it does offer additional competition, but it does not remove the opportunity for knowledgeable professionals to buy items which their expertise indicates are good value and which the inexperienced will avoid due to the high risk of a dud.

Attendance at auctions solely for the purpose of obtaining information on the value of different types and styles of clocks is entirely fair. Another method is to attend pre-sale viewings of similar types of items and study the pre-sale estimates in the catalogue. Alternatively some houses issue lists of sale prices achieved, at a small fee. Rural sales are often reported in the local press.

The other route to understanding current values is to visit the high-street dealers, or antique and specialized fairs. Take the opportunity to ask about the type of clocks which are of interest to you, even if none are on offer for sale. There may be something in the dealer's restoration workshop which will soon be for sale. If selling, they may have a customer seeking to buy your type of clock.

It may be possible to obtain advice from a known restorer or dealer before purchasing. This may incur a charge but can be especially helpful if an outline estimate of the likely cost to restore or repair the clock is included.

It may be advantageous to obtain a proper valuation of the clock from a suitably qualified valuer or expert prior to sale. This is very different from asking the same person what he or she is prepared to offer for the item. The valuer is quite likely to make a charge for the effort, such as the research required to provide a realistic valuation. Prior to making this estimate it is essential to specify the basis on which the valuation is required. The valuations given for the purpose of purchase, sale, probate and insurance will all be based on different assumptions.

As a general guide, the most valuable clocks are those made by the most expert and best-known makers. The greater the amount of work in construction, the higher the value. In addition, a piece with a longer going period is usually more valuable, as is one with extra features such as automata, a musical movement or an astronomical calendar. Size and condition are two other important factors. A clock which is too tall will have a limited range of buyers unless it is special in some other way. The need for restoration, or poorly executed restoration, will always reduce the value.

3   Restoration and Repair

Restoration of a clock may be desirable or essential; it may simply improve the visual appeal, it may redress and prevent long-term deterioration. It should be carried out with a full knowledge of the long term impact.

The decision to undertake restoration on a valuable clock should be made on the basis of three factors: the historical interest of the piece, the financial value, and its external appearance. A professional restorer will be happy to offer advice in these areas. The greater the value and the historical interest in the clock, the more consideration needs to be given to this matter. The ultimate decision is clearly the owner's, who will need to choose between the following three courses:

- **to conserve** – to do just enough work to maintain the piece in its current original condition so that any further deterioration is minimized; this is most relevant to unique pieces or those of great historical interest

- **to restore** – to return the piece to a condition similar to that when it was made

- **to repair** – to bring the piece back into working order; this is most relevant to modern mass-produced items

In practice the course chosen may cover more than one of these options.

It is rarely of benefit in the long term to undertake restoration work which is of poor quality, which is outside the overall style and finish of the original piece, or which changes the originality beyond its conception. It is inevitable that some work will be necessary to counter damage and wear and tear. Nevertheless, the owner may need to decide whether it is best to restore fully or partially. The latter option may necessitate that the clock runs for short periods only. Clearly this type of situation is only relevant to some of the oldest and most historically interesting pieces. Nevertheless, where new technological developments have taken place in recent times, this approach may still be valid even for a relatively modern piece.

Good Practice in Restoration

To assist the professional restorer and to protect the public detailed guidance on restoration was issued by the British Horological Institute in 1995. This spelt out in detail what constitutes good practice when considering the repair and restoration of an antique or collectable clock. This guidance was drawn together by the most eminent horological experts in the UK. It recognizes that old clocks are as much a part of our heritage as oil paintings, ceramics and other works of art and there is a responsibility on all to keep them from the harm which can occur from inappropriate repair or restoration. It is desirable that owners are aware of this recommended practice so that in discussion with restorers they are in a position to make an informed decision on the best course of action.

The following is a summary of this recommended good practice.

1 When new components are made to replace ones that are

missing or beyond repair they should be made in the same style and finish as the original. New parts should be marked to identify them as replacements and the originals kept as part of the history of the clock. No attempt should be made to fake replacement parts.

2   Any additions or repairs to a clock should be completely reversible. No new holes should be made in a clock movement and any new components should be completely removable. The original surface finish should be retained.

3   Where parts such as escapements have been converted 'in antiquity' to a 'new' style, careful judgement is necessary before deciding whether to reconvert to the original style.

4   Where marriages of parts not originally together have taken place, objectionable parts can be removed and replaced.

5   When new components are made or repairs done to a component, the same materials as the original should be used.

6   The owner and the restorer owe a duty to the item to preserve it for future generations.

It is by adhering to these general principles that the national stock of antique and collectable clocks will, as far as practicable, be maintained for the benefit of future generations.

Cleaning of Clock Movements

The oil required to keep a clock movement operating satisfactorily slowly oxidizes, becomes thicker and forms corrosive deposits. After a few years, it becomes contaminated with particles of metal as a consequence of wear, and with particles of dust. The oil can spread to areas where it is not required, or worse, where it is harmful. If it comes into contact with the wheels and pinions, this can lead to rapid wear due to the abrasive action of the oil and dust.

Cleaning a clock movement is often essential as part of servicing and repair. The objective of cleaning is to remove old or congealed oil, corrosive deposits and metallic and abrasive dust particles. Clean, uncontaminated oil can then be applied.

There are different methods of cleaning the component parts of a movement, which differ in their severity. It has been common practice in the past, and still is with some restorers, to use ammonia-based cleaning agents. These are very effective at removing oil and dirt, as well as the dull coating on brass surfaces, resulting in clean bright components. The use of such cleaning agents has been a matter of debate for many years within the horological profession, however, because it can lead to inherent deterioration of the brass. For this reason their use is diminishing, although the risk of harmful effects is minimized by application of low concentrations for short periods.

Other methods can achieve the same essential objectives but do not necessarily result in a movement which appears bright and sparkling, so it is not valid to judge the thoroughness of the cleaning from the superficial brightness of the brass components. A different cleaning method may have been used. Alternatively, a clock with lacquered plates may have the original lacquering retained as a part of the originality of the clock, thereby preserving the colour and finish of the plates.

The signs of satisfactory cleaning and lubrication are the absence of dirty oil or other liquid or solid deposits, and the presence of clean oil in pivot holes and on rubbing surfaces. Wheels and pinions are in rolling, not sliding, contact and must be clean and oil-free. The only exceptions are likely to be where dark-coloured greases have been used for items such as mainsprings.

How to find a good Repairer or Restorer

A good repairer is one who carries out the work required to an acceptable standard, at an agreed price and delivers it by the agreed date. If only life was so simple!

Restoration and repair work on clocks is for skilled professionals. As with all work of a skilled nature, when seeking a restorer, it is advisable to seek a recommendation from a friend or contact who has already had similar work carried out. A good recommendation from a satisfied customer is a very good start. The next step is to pose additional questions to both the friend and the restorer. Is it the type of work which you are seeking? Will the work done be to the highest or lowest acceptable standard? Can the work be done in a reasonable time period? Is a guarantee provided for the work carried out? Does the repairer belong to a professional body? A satisfactory set of answers will give considerable confidence to the owner.

For those who have no friendly guidance and wish to seek possible repairers in their part of the country a special service is available, operated by the BHI. The Institute will give details of suitable qualified repairers in a particular area who are willing to be approached. The selection is based upon a profile of the type of work and special services which the member is able and willing to offer. Some of these members will never advertise their services – they have no need to do so, as their clients are all obtained by personal or professional recommendation. This is a free service, offered as part of the support of the BHI's members and is increasingly used by the public. Other bodies for professional restorers and repairers of antique and valuable items operate similar types of services.

The owner may enjoy cruising the countryside looking in the

premises of the various horological dealers, and making contacts and judgement on the restorer's aptitude. Members of the BHI will be proud to display their crest. To become a member of the BHI it is necessary either to pass the Institute's examinations or to demonstrate a high standard of practical work and considerable practical experience. Any person displaying the logo or the qualification of the BHI and practising professionally is bound to operate by the Institute's Code of Practice. Advertisements in local newspapers and trade guides are other sources for finding repairers. Local trade directories usually have a section listing qualified horologists who operate to a recognized code of practice. This guarantees the services of a qualified person and failure to carry out work to a professional standard can be referred to the BHI for judgement and rectification. Under this code, the member will:

- provide a written quotation or estimate for the work that is required; if further work is required then approval will be sought before going ahead

- submit an invoice detailing the work done when the item is returned

- have suitable insurance for the item, or indicate that insurance is the customer's responsibility

- not undertake work which is beyond his or her competence, and if specialist work has to be done by another professional, prior consent will be obtained

- only use the BHI qualification he or she is entitled to

- conduct the work to a professional and responsible standard
The Institute has its headquarters and museum at Upton Hall,

Upton, Newark, Notts, NG23 5TE, tel. 01636 813795/6. The Hall is open to visitors during many days of the year as well as for certain special events.

General Advice on Restoration

A detailed discussion of the work with the restorer is always fruitful before preparation of a formal quotation. Every effort should be made to speak to the person who will actually carry out the work, but it could be that certain aspects of the work will be contracted out. It is advisable to spell out to the restorer those aspects which do *not* have to be restored, and to discuss the standard of timekeeping to be expected from the restored clock. Ideas of accuracy may differ!

A good restorer may well have a number of items of work in hand at any one time with a queue of customers waiting for their work to be started, some of which will have been obtained on the condition that it can be completed quickly. It will therefore be necessary to negotiate the expected completion date. Once an agreement has been entered into a good restorer will make every effort to comply with this agreement to ensure continued goodwill.

A restorer of clock movements is unlikely to carry out the restoration of clock cases, unless the work is fairly minor. It is best to seek a specialist restorer for case work. This is particularly important where skills involving inlaying, marquetry, japanning or gilding are required.

It is to be expected that restoration work will not be obtained at a low cost. Good work can be expensive but it is the only way to ensure that the piece retains its full value. Competitive quotations for the work will ensure that a reasonable and fair cost is incurred.

It is beneficial to keep in touch with the restorer during the course of the work, particularly if the work is substantial. It will be more rewarding for both parties, and it will be easier to deal with any problems or questions as they arise and to ensure that the agreed completion date is met. A good restorer will allow the owner to control all decisions regarding the cost and the quality of the work.

The Most Common Causes of a Clock Stopping

The hands are bent so that they are catching each other, the dial or the glass.
The hands need straightening or pinning more tightly to retain in place.

The clock is lacking lubrication or the oil has congealed.
Cleaning and correct lubrication is required.*

The tick of the clock is very uneven.
The beat of the clock needs adjusting by either tilting the case slightly or adjusting the escapement.*

The clock is excessively lubricated.
The clock may need cleaning and correctly lubricating.*

The weights are catching on the case.
Either the case is tilting and needs correcting or the point of suspension of the weights is incorrect.*

The pendulum is rubbing on the case or part of the movement.
The case is tilting and needs correcting.

* *Seek specialist advice*
The chiming train is not wound up.

Wind it up.

The escapement has become damaged by moving the clock without first removing the pendulum.
Seek advice.*

The wear in the clock has become excessive.
Seek advice.*

The weight has dropped off the pulley.
Untangle the rope or chain and hang the weight on the pulley.

The mainspring has broken.
Seek advice.*

* *Seek specialist advice*

4 Originality

When viewing a clock with a mind to purchase, or when examining an existing acquisition, an owner is often faced with questions about its originality and whether it is complete in all its component parts. These are important matters and deserve serious consideration: they will undoubtedly have an influence on the value or cost of the item; they will have a bearing on the extent of any repair and restoration; they may limit the extent to which it can be restored to a satisfactory condition; and they will be important in determining whether it is a suitable item to purchase. From a collector's point of view, a clock which is original in all respects is always of greater interest, since it will fully reflect the style and materials of the particular period in which it was made. With few exceptions, a completely original clock will be of greater value than one which is not.

In the original manufacture of a clock the dial, movement, case and fittings will usually have been made in different workshops and assembled for sale. All of the components will have been made at about the same date, but with almost continuous use over many years, it is inevitable that parts may have become unserviceable, damaged, lost or out of fashion. Changes will, therefore, have been made to what was the original clock to keep it functioning. This is a never-ending process.

To further complicate matters, there is an additional present-day reason for another type of change. The increasing interest in and value of antique and collectable clocks has led to the deliberate assembly of clocks from old components. This is driven by the desire to create or improve a clock so that it has a higher potential value. It is clear that a small but increasing proportion of clocks which appear in the market place are assemblies of movements, dials and cases from other items. These are known as 'marriages'. Fortunately this practice is limited by the availability of the necessary components, but the owner or potential owner needs to be fully aware of this possibility.

When the lack of originality is the result of something which has already been restored, then the quality of the work carried out is clearly important. If the work was required to make good any wear and tear and has been executed entirely in keeping with the style and quality of the movement and case, it is probably an asset. However, if the work is poor in quality and out of keeping then it is likely to be to the detriment of the clock, its value and its collectability. It is frequently advantageous to buy a piece in an unrestored condition if the price is commensurate. This will allow good restoration work to be done without having to undo unsuitable work. The attitude of a professional restorer to the lack of originality in a clock will, in general, be to retain the status quo except where additions are objectionable.

There are also clocks in which the lack of originality is a substantial benefit. These are clocks into which a new or experimental feature has been incorporated. Such changes were made by those at the forefront of horological developments. Consequently, these clocks are of considerable historical interest and should always be preserved in their modified form.

What anyone examining a clock will quickly discover is that sometimes the evidence of its origins will be clear, but sometimes there will be considerable uncertainty. This is only to be expected. Many clocks will have been in service for over 250 years. Even with the same type of clock, different methods of construction will have been employed. A clock will often have been the only mechanically operating item in the household, and the owner will have meddled. Maintenance, repairs and sometimes updating will have occurred over the whole of this period in order to keep it in continuous service. Each clock will have become unique. The guidance offered below concentrates on those aspects which are most likely to have changed.

There are very many different types and qualities of clock. As a consequence of the scientific research into timekeeping and navigation needed to support a maritime nation, clock-making in the UK started to become an important activity at the end of the seventeenth century, and the manufacture of mechanical clocks, often of a high quality, continues today, even in the age of the electronic timekeeper. All mechanical clocks suffer at the hands of time in a different way and to a different extent. It is difficult, therefore, to provide detailed information on all possible changes to all the various types of clock cases and movements made over a period of 300 years. Nevertheless, there are many common features, and information on a change from originality in one type can be applied to a majority of the other types.

Examples are given in the photographs, to assist in understanding the main types of changes. These are drawn from the most commonly found types of clock but the same principles can be applied to those of the highest quality and greatest rarity. One important aspect must always be remembered: the older the clock the greater is the likelihood that changes have been made and the more extensive such changes are likely to be. Similarly, the younger the clock, and the lower its value, the less will be the need or motivation for change.

An understanding is needed of the possible reasons why changes came about and all the evidence needs to be gathered before reaching a judgement. When this poses difficulties, expert advice may be called for. Alternatively, further study and reading may be required on a particular topic to be able to come to a sound conclusion. There are many excellent sources of reference available and owners should have no difficulty in obtaining a deeper insight into their type of clock. It might even prove to be a voyage of delight and discovery.

The Case

The case of a clock provides a housing which will protect the movement from the dust in the air, from drops of moisture and from the risk of mechanical damage. Cases are, therefore, the first part to suffer wear and tear, particularly when in a harsh environment. Fortunately, clocks on mantel shelves, in padded carrying cases or on the wall are well protected from many of the possible risks. For this reason they are more likely to be in their original condition. Clocks at greatest risk are those standing on the floor and on wall brackets and these categories are considered in the most detail. However, many of the aspects discussed are clearly relevant to many of the other types of clock cases.

The Feet and Plinth

Feet which have stood on a floor for many years are the most vulnerable part of a clock. They will have suffered from animals, from vermin, and from human contact and most of all from damp conditions. It is very common in longcase clocks to find that the feet are completely missing or have been repaired or modified. Some clocks have no feet but do have a plinth. It is only by judging the overall style and proportions of the case and cases of similar age and origin that a judgement can be made on whether the feet are those originally fitted.

Early nineteenth-century English bracket clock showing one of the two brass brackets used to secure the movement to the case, with no sign of change from the original

A bracket clock bearing the name 'James Upjohn, London' in an ebonized case. The movement is sitting on an ebonized foot of different colour and finish to the rest of the case – an indicator that the case may well not be the original for the movement. Note that the movement is fixed in position by two bolts through the base of the case. Also note the vacant holes for the missing bracket for securing the pendulum

A Welsh early nineteenth-century longcase clock showing the correct method of fitting the seatboard. The board rests neatly on the cheeks of the trunk sides retained in this example by stout nails. This case has no runners fitted to prevent the hood tilting forward during its removal

Two longcase clocks which have suffered considerably in their lives. The nine-teenth-century case on the left has been cut down at the top and poorly repaired. The plinth has been shortened affecting the overall proportions of the case and a later moulding attached. The eighteenth-century case on the right has had the original arched top cut down and replaced with a flat top. The plinth has been considerably reduced and is out of keeping with the original proportions

It is rare to find that the feet have been replaced by copies of the originals, and once they are removed they are not usually replaced. A possible reason for this is to have a longcase clock which is slightly shorter to fit into a modern home.

Because of damp and decay, many plinths show partial and sometimes complete rebuilding. The dampness can spread easily from the floor upwards, causing deterioration in the lower 30 cm or more of the base. When judging whether reconstruction work has been carried out, an assessment of the style of each of the components is required. The original mouldings around the base, trunk and hood of a case will normally be of similar style, wood grain and shape. The mouldings will not, of course, be exactly the same shape, but will look in keeping with their place on the case. The strongest and biggest mouldings are generally around the top and bottom of the trunk. This is used as a means of blending in the difference in proportions of the trunk and the base and hood. Mouldings at the bottom of the base, and these could be single, double or triple mouldings, will usually be shallow to avoid them projecting excessively beyond the front of the base.

When considering reconstruction work it is beneficial to assess the correctness of the overall proportions of a case. The plinth of a longcase is usually a similar length to that of the hood. When the feet or plinth are removed or reduced, the shorter overall plinth looks out of proportion. Examples of the correct proportions for each type and style of clock can be found in some of the many excellent books on antique clocks.

It must however be recognized that local differences in style arise and there are no hard and fast rules on this matter. Books giving information on local makers and styles are available for some geographical areas.

A general examination of the type of wood and the direction of the grain in each component of a case is valuable. Each type of wood used will usually be of a similar colour throughout. Matching

A late eighteenth-century thirty-hour longcase clock where the feet (see detail overleaf) have been moved to provide a very plain case with a lower total height. It is an easy matter to restore the case to its correct proportions with very beneficial results

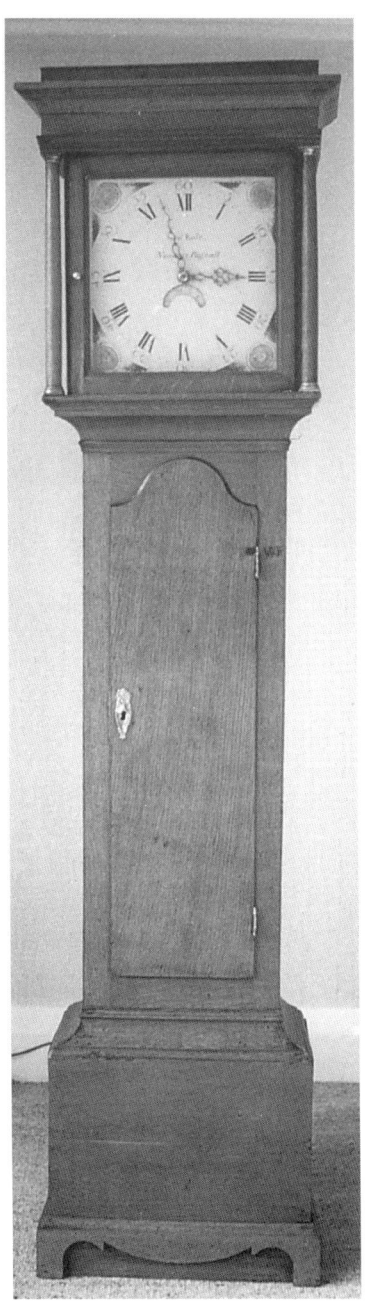

Detail of restored plinth of a late eighteenth-century thirty-hour longcase clock (see previous page)

parts would have been cut from the same or similar pieces of wood and the grain should be of a similar direction and cut. Significant departures from this suggest that a change from originality should be suspected and a more detailed search made for signs of a structural change or staining. Differences in wood colour are often very obvious except where serious attempts have been made to obtain a good match with the original.

The age and type of the case is an important factor. A country clock made in the eighteenth century, which has stood for many years on a stone cottage floor, is very likely to have suffered some deterioration. This is much less likely in a tall inlaid clock designed for the more spacious rooms of a Georgian or Victorian city businessman.

A common reason for repairs to a plinth is because a weight has fallen due to a gut or rope breakage. The damage can be severe, usually breaking the front bottom moulding, and this can cause bulging or fracture of the front panel. Where an internal floor is fitted inside the base, this can suffer damage, but the additional strength which such a floor can give may well help to protect the rest of the plinth.

When examining the plinth for signs of rebuilding it is necessary to look carefully inside – some form of illumination will be required. Differences in the type of wood used, areas where additional or replacement reinforcing blocks have been introduced and signs of the different types of glue which have been used will be easily visible. When the clock was made, the cabinetmaker would have used water-soluble glues, which leave characteristic runs of glue around some of the joining surfaces. Any wooden reinforcing blocks on the internal corners of the carcass would have been similar throughout the case. Tongue-and-groove joints were used in some better cases and nails were used in non-visible areas. Alternative forms of construction need to be examined in detail. Aspects to look for are newer or relocated reinforcing

A repair to the base of the plinth of an early nineteenth-century Yorkshire longcase clock showing the bottom moulding broken and the inlaid block panel cracked due to a falling weight. The feet are now missing and a later moulding has been added to strengthen the plinth

blocks, battens for strengthening, modern glues and the use of screws.

The front panel of the plinth or an applied panel (a block front), will usually be the widest panel in the clock case. As a consequence, it is likely to show the severest signs of shrinkage. Being restrained, the panel will either crack along the grain, open up a joint or become loose at one edge; each of these may have been restored to make good any damage. A sunken panel in a frame, which allows the panel to move without restraint, should show minimal signs of cracking.

Other aspects of the plinth needing possible repair are where veneer or other decorative items have become detached and lost, or where exposed corners are damaged. This can arise from many causes and such repairs are quite common, but if well restored they can be difficult to see. To identify replacement veneers it is necessary to look for differences in the direction of the grain and in colour. Replacement of thin strips, stringing, may necessitate the cleaning out of the gap, which often leads to a poor-quality repair. Also, it may be possible to see differences in level between the original and new pieces; this is easily visible if it is viewed with the light reflecting off it. Staining is often a tell-tale sign that some repair has been made since it is used to blend in replacement pieces of wood. Another sign that wood has been replaced is that the grain in the new piece is more open. (Colour illustration facing page 64, upper.) A good restorer will ensure that the grain is completely filled, leaving a mirror-type surface that is consistent with the rest of the case. Where these small repairs have been well executed they are often very hard to identify, and there should be no concern regarding their influence on the value or originality of the clock – they are just an inevitable result of proper care and restoration.

The feet on a bracket clock may be of turned or carved wood or cast mounts. These are often missing through damage or neglect. There will usually be marks or holes to show that feet were once fitted. Replacements are common but of little concern if in the style of the originals.

The Hood

Changes from originality to a hood can arise because of wear, damage or human interference.

The hood of a longcase clock will generally slide forward to allow it to be removed. Some of the oldest ones slide upwards. Wooden strips (runners) on the side of the trunk are normally fitted to allow the hood to slide safely, without falling forwards. These are occasionally worn, damaged or missing and are easily replaced. The weight of the hood bears onto the mouldings at the top of the trunk. With use, some wear will occur on the

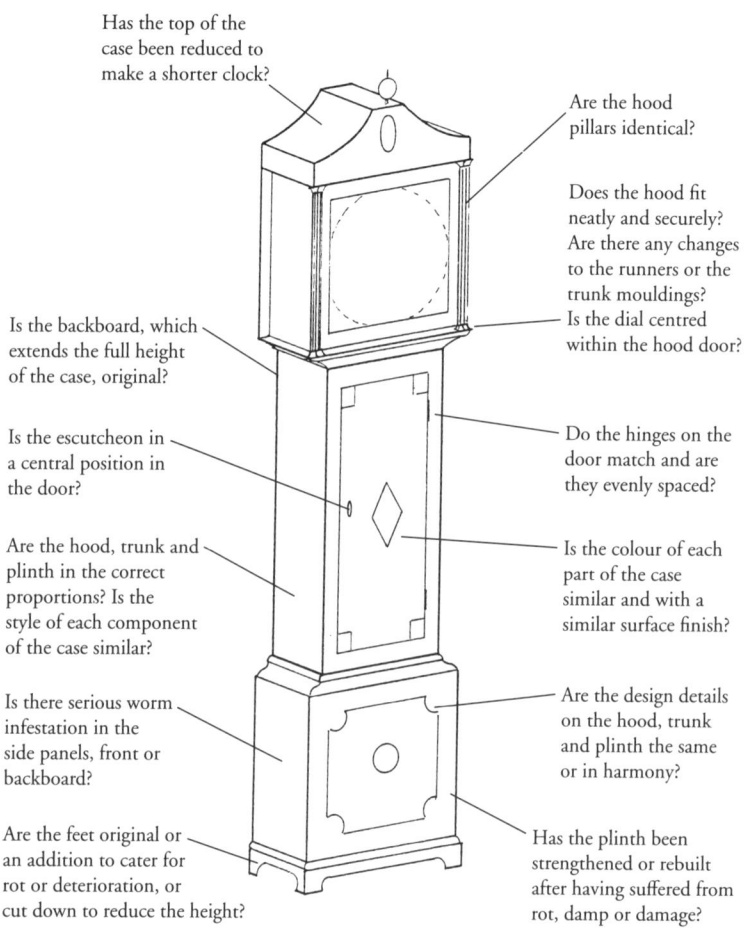

Has the top of the case been reduced to make a shorter clock?

Are the hood pillars identical?

Does the hood fit neatly and securely? Are there any changes to the runners or the trunk mouldings? Is the dial centred within the hood door?

Is the backboard, which extends the full height of the case, original?

Is the escutcheon in a central position in the door?

Do the hinges on the door match and are they evenly spaced?

Are the hood, trunk and plinth in the correct proportions? Is the style of each component of the case similar?

Is the colour of each part of the case similar and with a similar surface finish?

Is there serious worm infestation in the side panels, front or backboard?

Are the design details on the hood, trunk and plinth the same or in harmony?

Are the feet original or an addition to cater for rot or deterioration, or cut down to reduce the height?

Has the plinth been strengthened or rebuilt after having suffered from rot, damp or damage?

Fig. 8   Parts of a case which may be altered or not original

mouldings or on the underside of the hood, particularly at the front. This may make the hood sit a little lower, making the dial look a little too high. The evidence is usually easy to see and any wear is unlikely to have been repaired, due to the problems in doing so.

Longcases are sometimes reduced in size specifically to ensure that the clock will fit into the desired place. The finials are often missing for this reason, but it should be noted that the brass from these finials is often very thin and prone to corrosion. As a consequence, they start to fragment and look unsightly and may be removed for this reason. Of the two options available, reducing the height of the base is not the first choice, being difficult to accomplish. The easiest and most common remedy is to remove the top of the horns or other pediment, which is usually very obvious and unsightly. To counteract this barbarous act of decapitation it may have been restored at a later date.

In the unusual situation where the hood has been replaced with one of a similar age, probably due to terminal damage or loss of the original, this will have necessitated a change to the wooden runners, or to the supporting mouldings on the trunk, to make the hood door central with the dial. Steps will also have been necessary to make the hood the correct width and depth to match the dimensions of the trunk. Clear evidence will be left inside the hood or on the top of the trunk, even where the style of the hood is indistinguishable from the rest of the case. A complete replacement of the hood, or the hood door, should be easy to detect from the external appearance of the wood. It is very difficult to obtain the correct match of wood, grain, colour and polish with an old trunk and a new hood.

The glass in the hood door is prone to damage and is often replaced even when the door is original. The correct glass will be of the older type, containing ripples and bubbles. (Even when cracked,

An early nineteenth-century English longcase clock where the top of the 'broken' pediment has been removed. Even when restored to a good standard (opposite), there are still tell-tale joining lines left behind.

A restored pediment on an early nineteenth-century English
longcase clock

the old glass is often best retained provided that it is not a hazard.)
Old picture glass may have been used to replace a broken original,
with a satisfactory result.

Hoods are sometimes provided with a lock so that the door
cannot be opened without previously opening the trunk door, in
order to prevent interference with winding or timing. The tongue
of the lock is attached to the hood door, being secured from within
the case with a wooden bar or bolt. These locks are often missing
or incomplete, even in otherwise original cases.

A dome top to the case of a bracket clock, or one of the various
other shaped styles, is always more attractive than a plain top. This
has led to changes to the top, sometimes in previous centuries to
reflect a change in fashionable style and sometimes in recent years to
create a more attractive and valuable clock. The evidence of recent
changes should be easier to identify using the advice offered above.
Older changes will be much more difficult to spot, and expert advice
may be required, especially when considering a high-value clock.

Shaped tops on clocks are particularly prone to cracking in the hot and dry conditions found in the home. Evidence of reinforcing or restoration is not uncommon and ought not to be confused with changes for other reasons.

The Trunk

Of all of the parts of the clock, the trunk is the most likely to be completely original. It is reasonably protected from the most common sources of damage. The most likely change is to the door. It is prone to damage and may have been repaired, but it will rarely have been replaced. The hinges are subject to wear and tear and if replaced an identical style should have been used. When originally fitted the escutcheon for the key would be close to the middle of the door and the hinges would be nearly equidistant from each end. If the door has been replaced, there may be changes to these items. It is useful to examine whether the bolt of the lock, if present, engages with its mating hole in the side of the trunk. Locks are often missing and replaced with ugly wooden knobs or catches.

Doors are particularly prone to shrinkage because a wide plank of wood is used. Being unrestrained at the edges, it warps rather than cracks, which may lead to problems with neat fitting. Where this has become unsightly, an attempt may have been made to improve the situation. This could include refitting the hinges or fixing wooden supports onto the rear of the door. Severe cracking may have occurred where the method of construction has prevented normal shrinkage. Where a door is veneered, shrinkage will have caused a warping of the veneer, and where the door has cracked, the veneer will also have done so. This can be very unsightly and a professional repair will have been necessary to achieve a satisfactory appearance.

Originality

Shrinkage of the door can lead to the bolt of the lock no longer reaching its slot, preventing locking or proper closing. This can be irritating for the owner and changes may have been made to allow the door to be secured.

The trunk door hinges on this oak longcase are clearly original with no sign of changes in their position

Shrinkage can occur to the backboard of the case, as this is a restrained panel. This shrinkage is not usually too severe and does not normally require remedying. The whole or part of the backboard may also have been replaced, as being against the wall it is more liable to pick up dampness. The lower part is at greatest risk since the dampness in a wall will not rise more than a metre. The backboard is generally of softwood and woodworm is very common. Changes to the reinforcing blocks at the internal corners will be the most likely tell-tale sign of replacement, or a piece of wood of a different age or type. Old floor boards are a convenient material to use for replacement, being long and straight. Signs of tongue-and-grooved joints and nail holes are some of the signs to search for.

If the sides and front of the trunk are fully veneered, then the carcass material will probably be oak or pine. Pine and the sapwood of the oak are prone to woodworm and sometimes the worm will penetrate through the veneers. Fortunately, it is easily treated provided that the wood has sufficient strength still to be fit for service. Where worm has reduced the strength of the wood or it has crumbled away, attempts may have been made to strengthen it either by replacing damaged parts or by using battens glued to the surface.

The other two parts of the trunk which are commonly modified are the seatboard and the sides on which it is supported (the cheeks). It is the seatboard which directly supports the movement, and problems with these are described below under 'The Movement'.

Other Departures from Originality in Cases

Damage to wall, bracket and mantel clocks due to damp and decay is not common. A more likely cause of damage in these clocks is due to falling. This will have caused damage to the edges and corners and, in severe examples, the fracture of joints, panels and fittings. Evidence of these types of damage and their repair should be fairly obvious.

There are other ways in which cases can depart from originality. In a few instances cases, which are now antique, have been made to house an even older movement. Examples of Victorian, Edwardian and later cases are also found. These may be in the original style and deliberately stained and treated to hide their youth, or made in a modern style. Occasionally, a case may incorporate a small part of an earlier case such as a veneered trunk or door, where the quality of the part used is sufficient to warrant the cost and effort to rebuild.

A major change carried out in Victorian times was to embellish the whole of an original case with carving. This was normally done on plain oak cases, but it was not restricted to this wood. The carving was usually to a good standard, but not always. It was then stained and varnished in a dark colour.

Pine cases can appear in plain wood, polished or painted. Originally, these would have been overpainted and stained or decorated. In recent times, many have been completely stripped to meet the demand for light, clean wooden furniture.

If no specific signs of change from originality are found in a clock case, it is a useful exercise to make an overall assessment of whether all the parts are in a consistent style for the clock. The pillars on the hood should be similar to or sympathetic with any on the trunk. The corners of the trunk should be in sympathy with the corners of the base. Veneer and inlay should be used to the same extent and in the same manner on all of the front parts of the case. The sides of a longcase or wall clock are exceptions; they are normally not decorated in the same way as the front because they are not so immediately visible. In bracket and mantel clocks the sides are usually decorated in the same manner, although not so flamboyantly. When judging the case as a whole, all parts should be in keeping with each other in terms of colour, finish, decoration and attachments, except in those areas where deterioration has taken place for some plausible reason.

An examination is warranted into the items of decoration on the case. The brass or wooden finials on the top of the case can be removed, and may be lost, when a case is cut down. In some clocks the wooden or brass frets which let the sound of the striking or chiming out, are broken or lost. Handles and decorative carved or gilded cast mounts are some of the peripherals which can become detached and lost. It is so much better when these items are original but the ravages of time take their toll and replacement with items of similar age and type is the only recourse for the owner.

Staining

Brief mention has been made of the use of staining as a part of repair and restoration work. It is useful to consider this process in more detail, since it will make it easier for the prospective purchaser to recognize when it has been used. It was, of course, used by the original makers in some cases to give a different surface appearance. For example, some oak cases were stained to resemble more expensive woods such as mahogany. The insides of trunk and hood doors are occasionally found stained or polished, presumably to look better while briefly on view. Sometimes the central part of the backboard which is on view when the trunk door is opened is veneered to give a pleasing effect.

The main purpose of staining in repair and restoration work is to hide a new piece of wood so that the overall piece is of a consistent colour. It is also used to tone surfaces which have lost their colour, particularly where prolonged exposure to sunlight has caused bleaching.

Originality

Careful choice and blending of stains can give a finished colour to new wood which is a good match to the original. But it is difficult to maintain this match in both daylight and electric lighting, so if possible, compare a suspicious area under both lighting conditions. Another useful check is to look carefully at the grain of the wood. On the external surfaces of old wooden furniture the grain will often be full of the varnish and polish accumulated over the years giving a mirror-like surface. With a new piece the grain has to be filled correctly to achieve the same effect. A piece of wood with an open grain alongside an original piece with a completely full grain is suspect.

Surfaces on the inside of a case the wood will not have the accumulated layers of polish and should be matt in appearance, with an open grain. It was normal when a case was made for any colouring or finishing to be restricted to the outside surfaces. The inner surfaces will have been left completely untreated, except where stain or polish has run over the edges. Any stain on the inside of a case will not normally be original, except for those special situations mentioned above.

Another tell-tale sign of staining is revealed when the surface of the wood has been knocked or cut. This will show the colour of the underlying wood. In old wood, the colour will be the same all the way through since it will all have aged over the same period. When a wood of a different underlying colour can be seen, usually lighter, this is a sign that the piece of wood is not old, but made to look much older by a surface coating of stain or coloured varnish.

Ebonizing is a technique which was used to make a wooden item look as if it was made from ebony. This technique was used in furniture of all types as well as clock cases. It consists of a black coating applied to the surface, with a varnish or lacquer on top to give a shiny surface. Being a coating, the underlying base wood shows through where corners or edges have become rubbed or knocked.

This is usually a hard, fine-grained wood which is lighter in colour. This technique was also used on the inside of cases to give the same effect but without the shiny surface. Ebonizing is distinctive, and is unlikely to be confused with staining which has been used specifically to disguise a change or repair. Where ebonizing is repaired a good match of the black colour will usually have been obtained. A good match of the faded surface lustre is more difficult for the restorer to achieve and will be a useful indicator of a change from originality.

The presence of staining is often apparent because stain has a tendency to remove the natural inherent lustre of the wood, so that it looks dull and unattractive. A good restorer will match the colour, fill the grain and provide a surface lustre the same as the original. He will of course use a piece of wood of the same age and type as the original, if available, since this overcomes many of the potential problems of matching. When the work is of this quality, it will usually be an asset to the clock.

Metal, Ceramic and Marble Cases

A silver or gold case would be expected to have a hallmark. Unfortunately, these can become worn with regular cleaning and be indecipherable. The markings will normally be on all the separate parts of the case and will be on the body of each item, not on a removable label or panel. When present, the first check for originality is whether the date is consistent with the estimated date of the clock. Any changes from originality could arise from repairs of damage, or a marriage with a different movement.

Base- and precious-metal cases will have considerable mechanical strength and therefore be resistant to damage, but, bending out of shape is more likely. Cast materials such as iron or brass are more brittle and with these fractures may have occurred and necessitated gluing or joining.

Ceramic or marble cases are unlikely to have been deliberately changed from originality. The most likely occurrence is that repairs will have been made because of accidental damage and this will have resulted in a loss of some part or detail. Component parts may also have been removed where the original was considered to be too large. Glass panels in cases such as those of carriage clocks may have been replaced. All glasses would be expected to be of the same colour and a good fit in their respective apertures.

The Movement

It is not uncommon in longcase clocks for the movement not to be original to the case and for the dial not to be the original one fitted to the movement. This is less likely in other types of clock but is not unknown. It is highly desirable for the prospective buyer to be able to judge whether either of these changes has occurred. By careful observation, the majority of these marriages can be identified. The existing owner will also wish to look for signs of changes, although the discovery of a hasty marriage may be disconcerting.

Is the Dial Original to the Case?

The first thing to do is to assess whether the dial fits into the case correctly. When the clock was made the dial would have fitted neatly into the aperture of the case in order to be attractive as well as to keep out the dust. Any small gap between dial and frame should be equal all around. Sometimes there is no gap because the woodwork (the dial frame) or bezel overlaps the dial slightly. In such cases the degree of overlap should again be equal.

In some wall and mantel clocks the dial is masked by the door of the case. The aperture may be square, arched, circular or semicircular, but the same considerations will apply whatever the shape. A dial which does not fit centrally and neatly into the aperture is an indication of a change from originality.

Where brass spandrels or other items are attached to the dial these should be fully on view and in no way obscured by the edges of the dial frame. If the hood is original, a dial which does not meet these requirements is usually a replacement, but before making a judgement it is useful to check that the hood is sitting correctly on its supports. Serious wear on either the bottom of the hood or the top of the mouldings in contact with it can easily give the erroneous impression that the dial is too high. Similarly with doors on the front of wall and mantel clocks. The hinges should hold the door in its correct position.

Dials in longcase and many other types of clock were made in a series of standard sizes. If the above checks show that the dial is of the correct size, it does not necessarily prove that it is the original; it could be one of the same size. Further investigation is therefore needed.

Another initial check to make is whether the movement, which should be sitting on or attached to a seatboard or bracket, is temporarily displaced sideways. In this case, the height of the dial will appear to be correct, but there will be a bigger gap on one side than the other. If practicable, it is necessary to move the movement or seatboard until the dial is correctly centred. It should then be possible to determine whether it correctly fits the aperture side for side. It is often easier to make these checks when the dial is a break-arch or circular type, since the whole perimeter of the dial has to fit consistently for it to look correct. (Colour illustration facing page 64, lower.)

If the dial does fit correctly, although it is not proof that it is original, it is a good omen for further examination. One technique sometimes used by makers was to form a shallow recess or ledge in the front of the trunk into which the lower part of the dial fitted snugly, making it less likely to move. When present, this is a very useful additional guide as to whether the dial is of the original size.

If a change of dial is suspected, it is necessary to examine whether the dial frame or bezel in the hood has been made slightly larger to accommodate a larger replacement dial. When originally made, the opening in the frame would have been slightly smaller than the glazed aperture in the hood door to allow the whole of the dial to be seen without obstruction. Any change should be obvious from the colour of wood around the aperture where it has been cut, but staining may have been used to camouflage this. A useful tell-tale sign in a hooded case is a small horizontal slot cut into the lower left-hand part of the dial frame. (Colour illustration facing page 65, upper.) This slot is to engage the catch on the hood door. If the opening in the mask has been made larger, it is likely that this slot has been partly cut away or is now too close to the edge of the mask.

Where a smaller dial has been inserted, a large gap around it will be obvious unless some action has been taken. To fill in this gap a narrow shaped moulding may have been fitted around the perimeter of the dial. Alternatively, the complete dial frame may have been replaced with one to suit the new dial size. Such a change is usually obvious from the style and type of wood used.

A rare type of change in a longcase clock is to make a square dial, with its original movement, fit into a case intended for an arched dial. To do this, it is necessary to add an arch to the top of the dial. This will be a half- or part-circle in brass or painted iron, to match the original dial. Alternatively, it may be a circular rim to allow a moon dial to be installed. These arches will be screwed or riveted to the original dial. A careful examination needs to be made of such changes, particularly where the added part is provided with engraved scales or decoration, since these extension pieces may be the ones originally fitted.

The aperture in the wooden mask may have been made smaller using a beading

Or increased by cutting away to accommodate a dial of a different size

The bezel will allow the minute markings to be fully in view with no gap between dial and bezel

The mask may have been completely replaced to fit a different sized dial

The aperture in the hood door will be slightly larger than the dial mask allowing the dial to be fully in view – the minute circle will be unobscured. Changes to originality may infringe this aspect

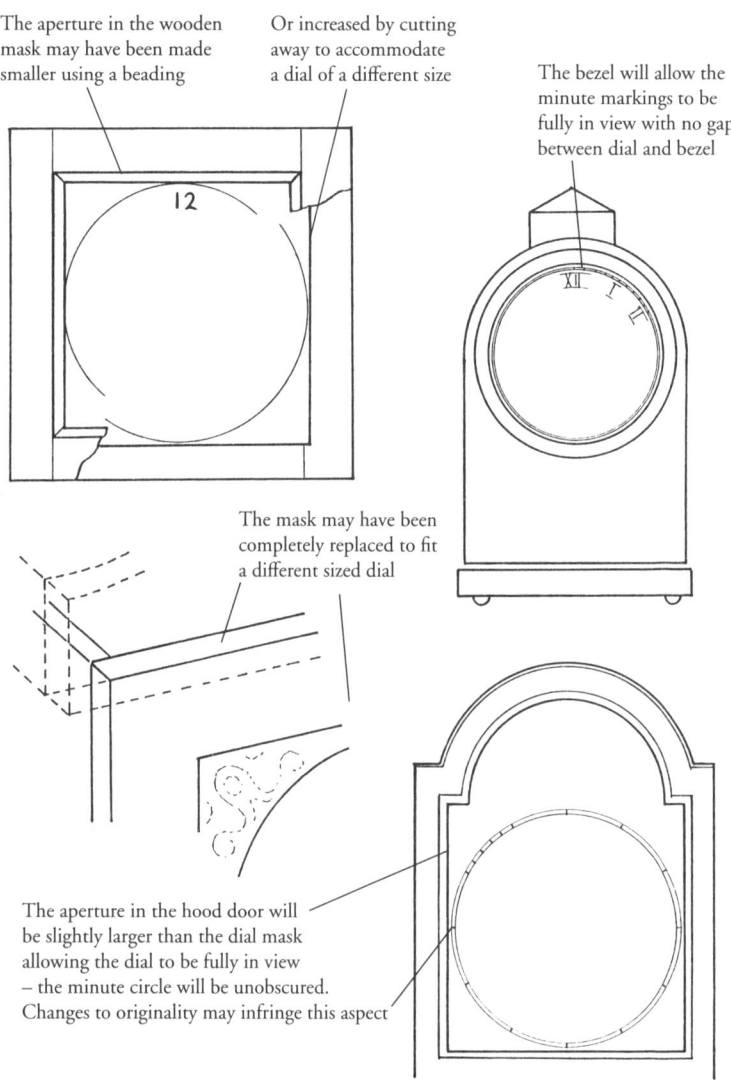

Fig. 9   Possible changes to a hood or case to fit a replacement dial

Example of a good quality late eighteenth-century eight-day brass dial from a clock by William Latham of Macclesfield which shows evidence that an arch was once fitted to the top by three screws

An assessment of originality should include whether the dial is of a similar date of manufacture to the case and movement. To make this judgement, the owner will need to gain a knowledge of the changes in style over the years. Fortunately, these have been well researched and there are some excellent books on this topic.

One reason for changing a dial is that serious damage may have occurred at some time. This is not common, since a damaged dial still serves its purpose and repainting can usually provide a satisfactory repair. Enamelled dials are much more difficult to repair and are likely to have been left cracked or damaged, unless they are very unsightly or unserviceable.

Another reason for a change is that the dials from two clocks may have been exchanged. This may seem a very odd thing to do, but the objective was to fit the more valuable of the two dials to the more valuable of the two movements, in order to increase the value of the two clocks. The changeover would generally be achieved by fitting the eight-day movement of a painted-dial longcase clock to the brass dial of a thirty-hour clock. Two winding holes will have been required in the brass dial. This marriage leaves the eight-day painted dial with the thirty-hour movement and where this combination is found it could well be for this reason. Fortunately, these changes are no longer financially attractive. They were popular when the value in the market of eight-day brass-dial clocks was well in excess of those for either thirty-hour or painted-dial eight-day clocks.

When examining a dial it is necessary to be on the lookout for those which appear to offer more than the movement can provide. A dial with two winding holes, which looks as though it is on an eight-day clock, can be fitted with original but false winding squares and an original thirty-hour movement. This type of clock was made specifically to deceive and to impress the neighbours and friends! There are in circulation clocks with dials which have a painted moon wheel in the arch of the dial, which is solid and not intended to operate. Dials with false painted winding holes and squares were produced by some makers. Much rarer are dials with a 'strike/silent' hand in the arch of a dial, which is unused. These were all made deliberately to mislead the casual observer.

These clocks with false features are fairly rare, but they are interesting and collectable items and should not be mistaken for any kind of marriage. (Colour illustration facing page 65, lower.)

Having considered changes to the dial as a whole it is worth considering changes to its main components.

**FALSE FEATURES ARE INTENDED
TO DECEIVE THE OBSERVER INTO
BELIEVING THE CLOCK HAS
ADDITIONAL DESIRABLE FEATURES**

A strike/silent hand
may not be intended
to operate since no
striking train is fitted

The moon dial may be
an imitation, not capable
of operating, yet painted
to a high standard

A date aperture may
be a painted replica

Winding squares and ferrules may be
painted replicas

Winding squares may
be false consisting of
short posts mounted
on the false plate

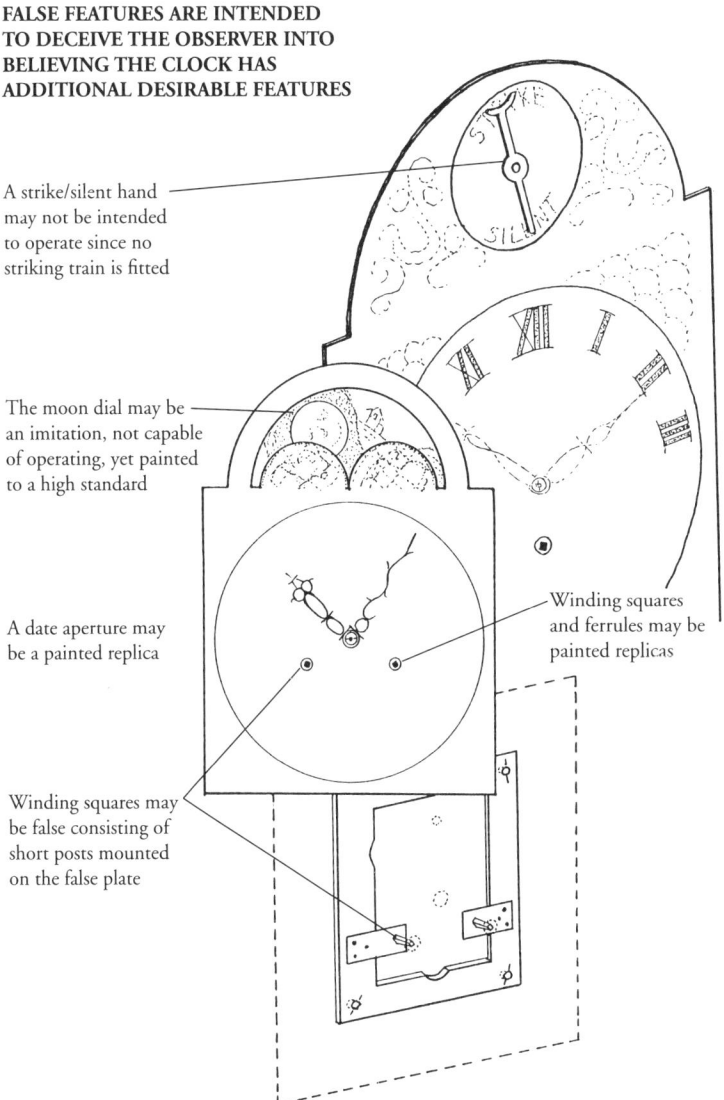

Fig. 10   Various false features in clocks

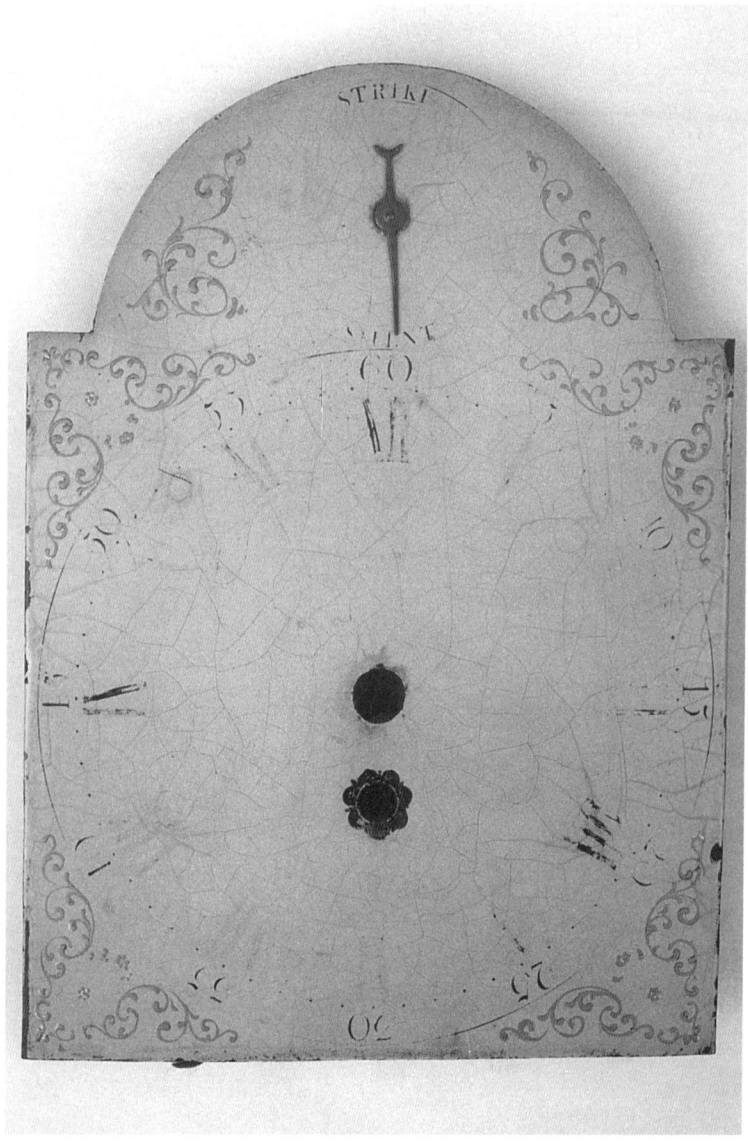

Example of an eighteenth-century English bracket clock dial which has a 'false' feature. The clock has only one winding square and therefore no striking train. The strike/silent hand has no useful purpose

Originality

A brass dial plate will often, but not by any means always, have a detachable chapter ring. Even though the plate may be original the ring may have been exchanged for another one. This could be for legitimate reasons such as damage, but is more likely to be a marriage, either to make the clock look older than the original or to provide a chapter ring with a different or more prestigious name engraved on it. The fixing of the chapter ring to the dial is by three or four short feet, with steel or brass pins. Evidence of new feet, or new or moved holes, should be sought. The rear of a brass dial and chapter ring should show the original hammer (planishing) marks consistent with its original manufacture and the marks left by the insertion of the various retaining pins. Other marks will need careful examination for signs of change.

The dial plate and chapter ring in a brass-dial clock will each have been originally cut from sheets of cast and hammered brass. Sometimes the front surface of the dial plate may be gilded and the chapter and any other dial rings silvered. Although these are good signs, neither is a guarantee of originality.

The dial plate is usually, but not always, cut away behind the chapter ring in a series of sectors to avoid the waste of valuable brass. The roughly cut edges of these sectors are usually clear to see on an original plate. The rear surface will often be covered with dirt and a hard layer of verdigris, but not with paint or varnish in an attempt to hide a change from originality. If evidence is found on the rear of a brass dial of part of an engraved design, this should be examined in detail. In rare discoveries original brass dials are known to have been made from earlier dials or from plates used for engraving practice. It is also wise to check that the dial is of solid brass, as a restorer may have used sandpaper sprayed with gold paint to provide a fake matted centre.

A more substantial change from originality is when both the dial and the movement have been changed, either with both being original to each other or from different clocks. To clarify whether either of these has happened, two matters need to be examined: the fitting of the movement in the case and the matching of dial and the movement.

Is the Movement Original to the Case?

To assess the originality of the movement to the case it is necessary to look carefully at the seating or fixing of the movement. Gaining access to see the inside of a bracket or mantel clock case is usually easy and visibility is not normally a problem since there is a generous space for the movement. This is also true in longcase clocks once access has been gained, but the complete removal of the hood is necessary. Wall clocks may be easy to access, but in some partial dismantling is required.

In a bracket clock, the movement usually sits on a loose or removable wooden foot, similar to a small narrow table, or as a shelf, supported on the sides of the case. These were originally made to ensure that the movement was at exactly the correct height so that the dial was perfectly centred in the dial mask or bezel. The layout of a movement is individual to a maker. The height of the centre arbor on which the hands are attached will therefore be at a set distance from the lower edge of the movement. Unless the movement is of identical height a replacement movement will require some alteration to the foot or shelf or their supports, in order to bring the centre arbor exactly to the centre of the dial. Signs of trimming to make a foot or shelf lower, or extra blocks to make it higher, should be sought. The foot could, of course, be a complete replacement. The original will usually have been in a soft wood or oak, as expensive hardwood would not have been wasted on these internal fittings. A replacement will be in a wood which is different from the original carcass or disguised by the presence of staining. Suspicion is warranted if either of these is found. Further, the original foot would be expected to have absorbed many drops of oil in its life. This all adds to the authenticity.

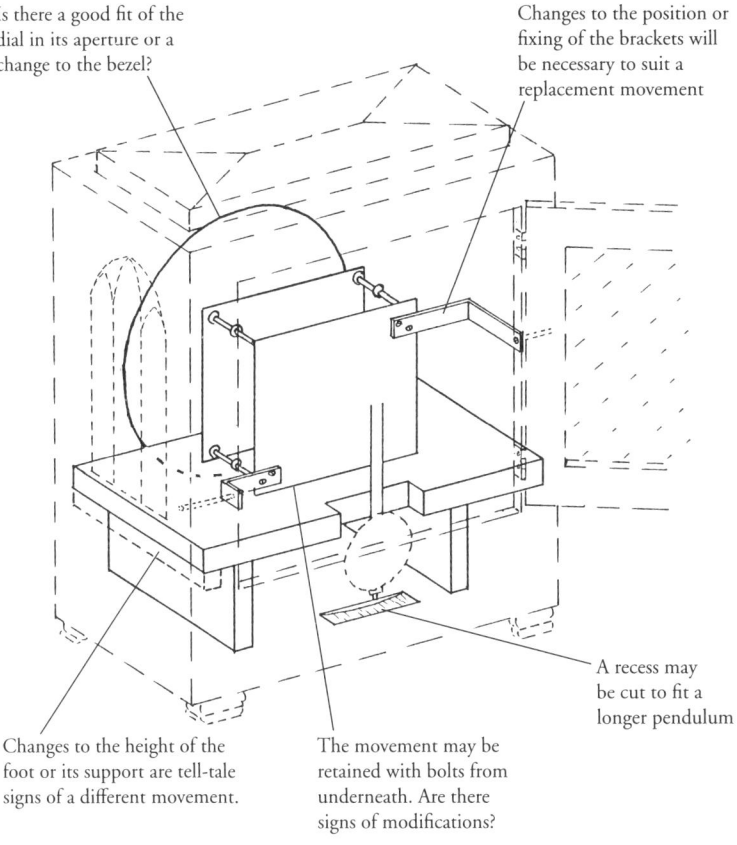

Is there a good fit of the dial in its aperture or a change to the bezel?

Changes to the position or fixing of the brackets will be necessary to suit a replacement movement

A recess may be cut to fit a longer pendulum

Changes to the height of the foot or its support are tell-tale signs of a different movement.

The movement may be retained with bolts from underneath. Are there signs of modifications?

Fig. 11   Possible changes to a bracket clock case to accommodate a different movement

The pendulum on this mahogany cased bracket clock is locked in position for transporting. The special screw is 'parked' on the back plate when not in use. The foot supporting the movement ensures that it is at the exact height for the case. There are no signs of changes inside the case

Originality

A second check in a bracket clock is to look at the method of retention of the movement. Different means of attachment have been employed. In all examples, the fixing has to be firm and strong because the movements are very heavy. Some very early bracket clocks use two or more latches on the dial mating with slots in the case. In later clocks two stout brass brackets were commonly used, which screw onto both the movement and the sides or back of the case. (Colour illustration facing page 80.) Other means employed include two long bolts between the movement and the base of the case, which will pass through the lower horizontal pillars between the plates of the movement. Relocation of the bolts or brackets strongly suggests a replacement movement.

A common method of attachment is using two screwed rods hooked into the lower edge of the movement plates, the rods being retained with securing nuts applied from beneath the foot or shelf. Where the fixing is only to the foot or shelf, this must also be rigidly fixed to ensure security of the movement. Other methods of attachment may be found. Changes in the position of these fixings in an original foot or shelf suggests a replacement movement. (Colour illustration facing page 81 upper.)

Other unused holes in the case will need careful study, but any in the base may be entirely original and a sign that the clock was previously fixed to a wall bracket for reasons of security.

Another option in a bracket clock is that a movement has been replaced with one of the same size from the same maker. This is possible since at times they were made in standard sizes, but it is unlikely and very difficult to identify.

In a longcase or other weight-driven clock such as a wall or tavern clock, the evidence of a change in movement is similar. When the dial is correctly positioned the hole for the centre arbor (on which the hands are fixed) is exactly in the centre of the dial mask. The centre arbor of the movement has to mate exactly with this hole. Again, since movements by different makers are slightly different, the distance between the base edges of the movement and the centre arbor varies. A change of movement therefore requires a change in its height or support.

The movement sits on a seatboard and when original will normally be found lying horizontally on top of the sides (the cheeks) of the case, probably secured by either nails or screws. Slots will be found cut through the seatboard to allow the ropes, gut or chain to hang freely below the movement. (Colour illustration facing page 81, lower.)

Changes to the position of the movement may have been accomplished by one of eight methods.

1 The seatboard may have been raised by inserting packing pieces.

2 The seatboard may have been lowered by cutting it away at the ends.

3 The seatboard may have been replaced by one which is either thicker or thinner.

4 The cheeks supporting the seatboard may have been cut away.

5 The cheeks supporting the seatboard may have been built up.

6 Packing pieces may have been inserted between the movement and the seatboard.

7 The bottom edge of the movement may have been cut off.

8 The plates of the movement may have been extended by the addition of extension plates.

The first six methods are all easy to carry out. The seventh is rare since there is usually little scope for it. And the eighth is an arduous task. But each leaves ample evidence. Nevertheless, it is

Originality

useful to have a little prior experience of the quality and accuracy of the fitting of the movement and seatboard in a completely original clock.

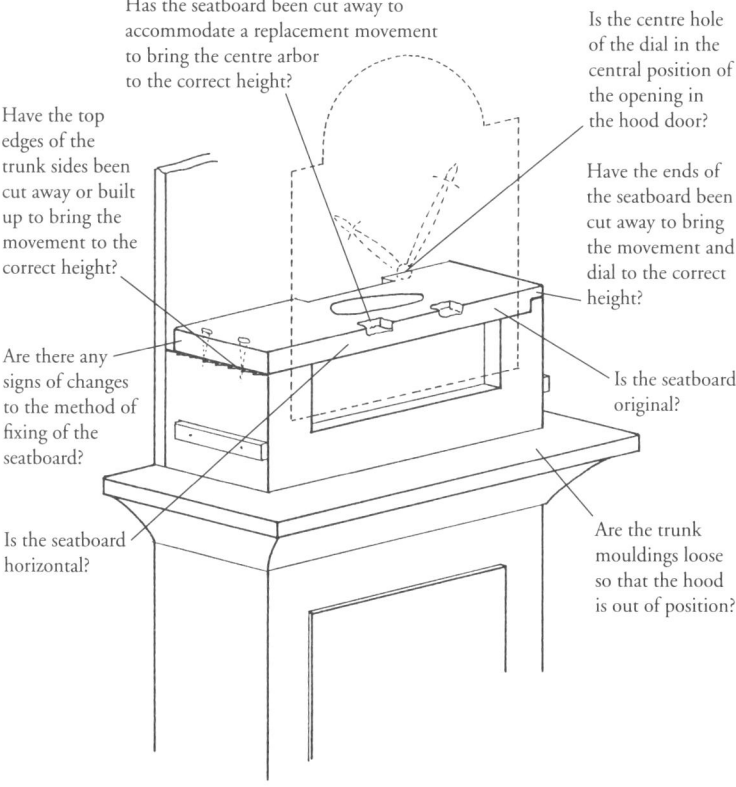

Has the seatboard been cut away to accommodate a replacement movement to bring the centre arbor to the correct height?

Is the centre hole of the dial in the central position of the opening in the hood door?

Have the top edges of the trunk sides been cut away or built up to bring the movement to the correct height?

Have the ends of the seatboard been cut away to bring the movement and dial to the correct height?

Are there any signs of changes to the method of fixing of the seatboard?

Is the seatboard original?

Is the seatboard horizontal?

Are the trunk mouldings loose so that the hood is out of position?

Fig. 12   Possible changes to a case to fit a different movement

Corroborating evidence of a change to the seatboard is the presence of spare holes in it. This may indicate a change in the location of the fixing screws or nails, but these may have been made when a clock was reinstalled or the fixing became insecure. Good illumination helps, because the evidence of these possibilities is usually plain to see when looking in the correct place. Unfortunately this type of evidence is not always conclusive since a change of seatboard could have occurred due to failure of the original.

The seatboard is a source of additional information, in the slots cut into it to allow clearance for the rope, chain or gut. In a thirty-hour longcase clock, there is normally one long slot cut into the seatboard, wide enough to allow the chain or rope with the pulley and counterweight to be dropped through. In an eight-day clock, there are two rectangular openings to ensure that the guts for the two weights hang clear of the seatboard for the whole width of the winding barrel. If a thirty-hour clock is found sitting on an original seatboard intended for an eight-day clock, it is likely that the movement has been changed.

One additional aspect to examine is the distance between the rearmost part of the back cock of the movement (on which the pendulum hangs) and the case. There should be a reasonable clearance to ensure that the pendulum does not interfere with the back of the case. The clock maker will have ensured that this was achieved, and too small a clearance may indicate a replacement movement. A similar conclusion could be drawn if clearance is provided for the pendulum by having the movement sitting too far forwards. This will be apparent because the hood will not slide fully back into its correct position.

It is possible that the movement may have been replaced by a much later one or by one which is not of the same type as was originally fitted. In the case of a modern movement this is usually all too obvious. Modern movements have thinner plates and wheels,

The replacement seatboard in this mid eighteenth-century thirty-hour clock has been set below the top of the sides of the trunk (the cheeks) to bring the non-original movement down to the correct height. This is a clear sign of a change to accommodate a different movement

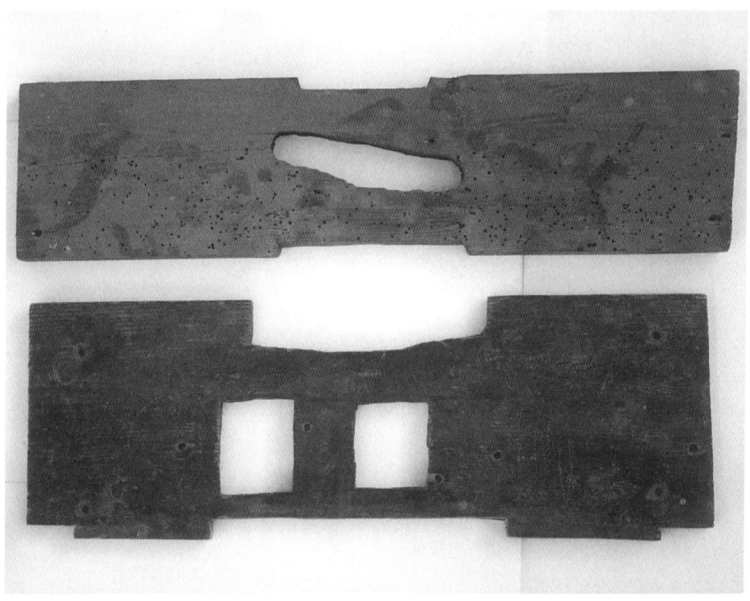

Seatboards from two longcase clocks. The upper example is from a thirty-hour clock whilst the lower example is from an eight-day

and are generally of less solid construction. They are also often skeletonized. It is only by looking at clocks of a similar age and type, however, that the owner will be able to identify a replacement by another old movement. One useful check is to judge whether the movement appears to be the correct size for the space available. Cases were made to house the movement and dial comfortably. Excessive or inadequate space in the case may suggest a marriage.

An examination of the top surface of the seatboard will provide useful information. A distinctive mark is usually present where the original movement has been sitting with heavy weights suspended from it for a considerable period. A slight depression in the wood will have been formed by the edges of the plates. If these depressions do not match the movement a change of seatboard or movement should be suspected. In rare examples, two or more small metal plates are attached to the seatboard to locate the movement accurately, which can be a good sign of originality.

The movement may have been replaced by a modern replica of the old movement. This is a difficult change for the amateur to identify, but is fortunately very rare. It may be more frequently seen, however, as older movements become less available.

Wooden-framed movements are occasionally used to replace the original metal-plated or framed movement. Similar types of tell-tale signs of such a change will be present as are found with plated movements.

When suspecting a change of movement it is advisable to examine the length of the pendulum. A different movement may require a different length of pendulum, which could necessitate a change to the space available in the case. If a longer pendulum has been fitted to a bracket or mantel clock, there may well be evidence that the base of the case has been hollowed out to accommodate the extra length. In addition, the original pendulum may well have left marks on the back of the case where it has occasionally rubbed. The marks can arise from contact with the bob or the rating nut; where they are not consistent with the pendulum fitted, a change from originality is the reason.

The weights fitted to a clock may have left distinctive marks on the case. Rubbing by the weights of a longcase clock at the bottom of the door aperture, leaving rounded hollows, is commonly found. The presence of two similar marks indicates that an eight-day movement should be fitted. A single mark suggests a thirty-hour movement. The only complicating factor is where the counterweight in a thirty-hour clock has also left a mark but to serve its purpose it has to be much lighter than a driving weight and any mark will be much less severe.

Marks have been made on the rear of the case where the centre of the pendulum bob makes contact with it, suggesting that the movement and pendulum have been together for some time and probably belong to the case

Pair of similar grooves at the base of the door opening showing that two weights were originally present and confirming that an eight-day movement was fitted into the case

Originality

In many mantel clocks, particularly those with circular-plated movements, the movement is attached to the dial and bezel and can only be removed from the front of the case as a complete unit. Access to these clocks for examination can be troublesome. Visibility into the rear opening into the substantial cases, which are often in black marble, is limited. Complete removal of the movement with bezel is required for a proper examination. The two long bolts in the retaining straps which also retain the rear cover have to be removed. Fortunately, removal is not usually necessary but it is beneficial if practicable. These movements are of a particular diameter and fit snugly. The diameter of the bezel is such that, when inserted from the front, it rests against the face of the aperture in the case. A complete change of movement requires one of similar diameter or a change to the bezel or case aperture. A replacement unit of the same size will be difficult to detect. A smaller movement will be unlikely to look correct and may not be tightly retained in the case. Changes to the length of the two fixing straps to make the movement fit into a case which is of a different depth is another possible sign of a marriage of a case and movement.

Mantel and wall clocks were mass produced at the end of the nineteenth century in both Europe and the USA. These movements can be of good quality, like the French round-plated movements, but many are of a lower standard. The high-quality French movements are fairly easy to recognize from the well-finished backplate with deep oil sinks, the prominent maker's stamp of quality, and the Brocot-type suspension, from which the pendulum is suspended, which enables the owner to make fine adjustments from the front of the dial. These movements were fitted into a very wide range of types and styles of mantel clock, some into cases of exceptional quality. It is to be expected that these clocks are much less likely to have been subject to the changes or marriages more commonly found in older clocks; where changes have been made this will have been due to damage or severe neglect but not to prolonged wear. With the lower-quality movements, such as those with skeletonized plates, wear will be more evident but replacement movements are not common.

Access to some of the lower-quality wall clocks can be a problem. The dial is attached to its wooden surround and is independent of the movement. The movement is normally attached to the back-board of the case; the fixing method favoured by American makers was with two or three wooden blocks screwed to the case. Removal of the hands as well as the dial with its surround is necessary to view the movement. Any changes of movement or dial will be evident from the points of attachment.

A similar method of assembly is found in some of the lower-quality mantel clocks, but other methods were also employed. The movement can be found screwed onto the inside of the front of the case, with the dial fastened onto the front, thereby allowing access to the clock through the rear door. It is easier to exchange the dial or the movement when they are independently fixed. Evidence of a change of movement may be the presence of new or unused holes in the case for retaining the movement, but if the original wooden blocks have been used there may be none.

Is the Dial Original to the Movement?

Having made a judgement as to whether the movement is original to the case, it is then necessary to attempt to determine if the dial is the one originally fitted to the movement. If it is possible to conclude that the movement is original but the dial does not fit the case correctly, the answer is already clear. However, the matter may not be that simple and it is necessary to examine in detail the fit of the dial to the movement.

The movement of this mantel clock is screwed to the inside of the case and the fixing is independent of the means of attachment of the dial. There is no evidence of any alterations from the original fixing

When originally assembled, the dial and the movement will have fitted almost perfectly, as this is easy to accomplish as a part of manufacture. A good clock maker will have achieved perfection. There will be occasional instances where old stocks of dials or movements will have been used and some type of modification will have been required, but these should be regarded as the horological curiosities that they are. Leave them for those who delight in the rare and unusual; they are unlikely to command a premium for their individuality. The task of making a replacement dial match a movement is more difficult.

For a replacement movement to be fitted to a dial, or vice versa, it is necessary to line up the various holes in the dial correctly with the mating parts of the movement. These are:

- the dial centre hole with the centre arbor

- the winding hole(s) with the winding square(s) (where fitted)

- the seconds-hand hole with the seconds arbor (where fitted)

- the feet on the dial, used to fix the movement to the dial, with the mating holes in the front plate of the movement

It is rare to find another movement which will perfectly satisfy all of these requirements, so a compromise has to be made. The outcome is a fit which is rarely perfect and usually an eyesore. Clear evidence is left.

Normally, but not always, the centre arbor will be located in the existing centre hole. This necessitates compromises in the other requirements. The first indication of a marriage is when the seconds hand (when fitted) is missing, because the arbor no longer lines up with the hole in the dial. It may just be lost, but it may well be the first sign of more substantial changes. The movement may not have the facility for a seconds hand, as is common in thirty-hour clocks. A dial provided with a seconds dial but with a movement without the facility for a seconds hand is a clear sign of a marriage.

To understand how changes are made it is necessary to understand in more detail how the movement and dial are assembled together. The dial is attached to the movement by a series of brass feet, which are permanently riveted at one end and slip through holes in the plate at the other. Pins, or in early clocks latches, are used for retention. This allows for secure fixing and easy removal. A false plate is often interposed between the dial and the movement. This is usual with eight-day movements with painted dials but not with thirty-hour painted or brass dials. There are exceptions to this rule. Experts have done research into this area and concluded that it was a preference of the dial-maker – dial-makers sometimes used false plates and sometimes did not.

Where a false plate is used two sets of brass feet are required. There are normally three or four in a set. One set are riveted to the false plate and the other set to the dial prior to the surface being finished. When originally fitted, the feet would have been solidly riveted and projected at right angles from the plates. They can become loose in normal service but rarely do so. Evidence that one or more are loose, such as missing paint or enamel or a repaired area on the dial, should be examined carefully. (Colour illustration facing page 160, upper.)

The existence of unused holes in the front plate of the movement or in the false plate is often a sign that a marriage has occurred, necessitating the use of differently located feet and maybe using some but not all of the original ones. The false plate only may have been replaced, but this is very unlikely. As a part

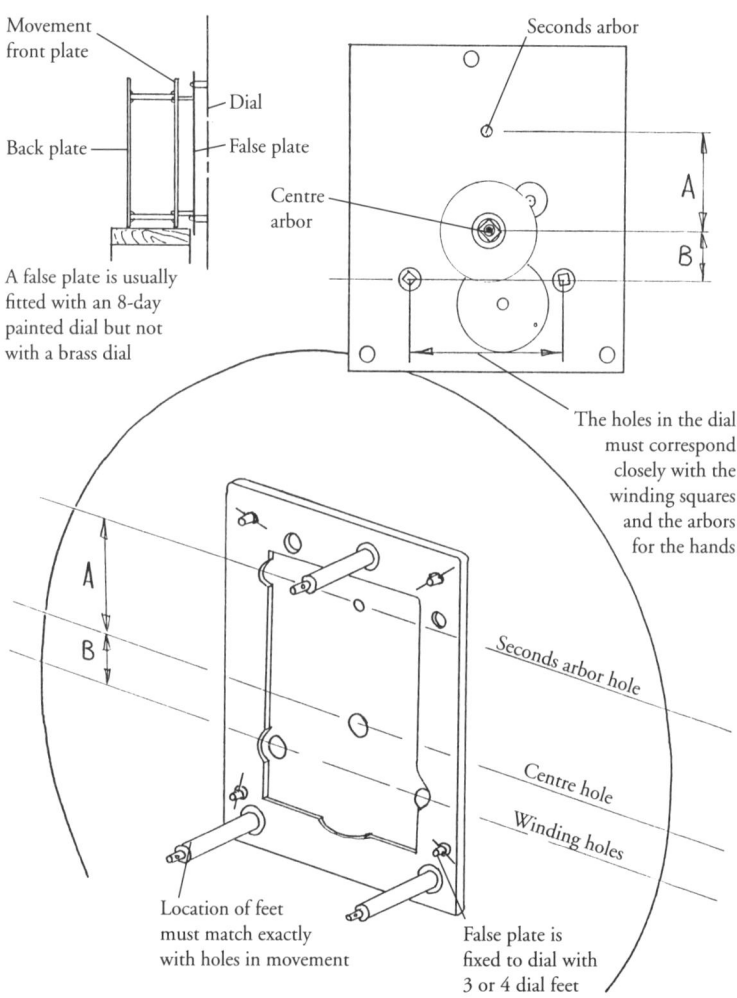

Movement front plate

Dial

Back plate

False plate

Centre arbor

A false plate is usually fitted with an 8-day painted dial but not with a brass dial

Seconds arbor

Centre arbor

A

B

The holes in the dial must correspond closely with the winding squares and the arbors for the hands

A

B

Seconds arbor hole

Centre hole

Winding holes

Location of feet must match exactly with holes in movement

False plate is fixed to dial with 3 or 4 dial feet

Fig. 13   Correct matching of movement with dial and false plate

of a marriage, an unused foot will usually be cut off but not completely removed, to reduce the risk of causing damage to the dial surface. An existing foot may also have been bent out of line. A newly made foot may not have been made to the original style or finish and will probably be clean and bright.

The seatboard is no longer horizontal due to the cheeks having been reduced. It has been cut away in two places at the front to allow the feet of the dial to be accommodated and thereby lower the height for the replacement movement. See other details overleaf

One of the four original dial feet at the top has been cut short since it would have interfered with other components parts of the replacement movement

The dial is now attached to the movement using three of the original dial feet with three additional plates riveted to the front plate. Since there are no unused holes in the front plate this movement has not previously been fitted to a clock – it was presumably unused

False plates were supplied by the maker along with the dial and are usually cast with the dial-maker's name. In longcase clocks and some wall clocks they were mainly of cast iron. Casting of metals used to be a fairly crude process. Pilot holes were cast in the false plate and the dial-maker or clock-maker filed the holes out to suit the dial feet. The holes in the false plate into which the dial feet are secured are therefore often not round or accurately located. Signs of changes to the fitting of the feet are consequently more difficult to judge and a detailed examination may have to be made. Some false plates were cut from sheet steel, this being the predominant type used in dial clocks, and these are usually found to be more accurately fitted. With this type of false plate, the maker's name (when present) will be stamped in.

A false plate fitted to a circular eight-day longcase dial showing correct fitting to the original dial. The holes in the false plate are normally cast in and are filed out by the dial or clock maker to fit the dial feet. The crude fitting is evident

In a clock with a brass dial, the feet were riveted through the main brass dial plate to which the other parts, i.e. chapter ring and spandrels, were attached. The surface was then filed and polished until the feet were not noticeable on the dial front. If the dial surface is decorated, the matting or engraving would have been done after the fixing of the feet. The design of the decoration would take no account of the location of the foot. If a new foot has been fitted it will show as an unengraved area, or if engraved, it will be in a different style or quality. If the riveting of the feet is hidden by the chapter ring, evidence of new or old feet will be visible only from the rear. New or different feet will usually be easy to identify.

A very crude form of marriage, already mentioned, is the fitting of an eight-day movement to the dial from a thirty-hour clock. As well as the necessary changes to the feet, this act of desecration requires two holes to be drilled in the dial for the winding squares. In a painted dial, this can be disguised from the front by the use of brass dial ferrules. With a brass dial, the holes will usually cut through the engraved decoration in an obvious or unsightly way. Occasionally, evidence is found of the reverse of this process, such as an eight-day dial fitted to either a thirty-hour movement or to an eight-day movement with differently spaced winding holes. This necessitates plugging up the two original winding holes in the dial.

The winding holes in the dial have been added or enlarged. This adaptation has cut through the engraved design to allow a different movement to be fitted. This is clear evidence of a marriage of dial and movement

An example of a late eighteenth-century brass dial from an eight-day clock in which the position of the winding holes is clearly completely original with the engraved design

On a brass dial the front surface will require careful engraving and matting to disguise the plugs. The plugs should be more obvious from the rear; less effort will have been made to hide the changes. On a painted dial of metal or wood, disguise is easier. A small area of repainting or a complete restoration will hide it completely.

A change of movement to a single-handed clock can often be found from a different type of evidence. Single-handed clocks were made up until the middle of the eighteenth century with brass dials. These are easily identified as single-handed clocks by the markings in the spaces between each hour, which are divided into four quarters.

Originality

No minute numerals or markings are provided, as none would have been required in a clock without a minute hand. Replacement of the original movement with a two-handed movement is, therefore, immediately apparent. The owner will have paid to have this work done to bring the clock up to 'modern standards'. It is a change which would not be considered by modern repairers or restorers; it is much more likely for the process to be reversed to create a more original and desirable clock. More of these reversals would be carried out today if there was a supply of suitable movements, but because these movements were worn out they have been largely discarded or used to repair other clocks. If the dial can be removed, it will usually be seen that the centre hole has been made larger. This is to accommodate the wider pipe for the motion work for the two hands.

This brass dial clock has unused holes on the rear of the dial plate which are covered from the front by the chapter ring. The holes were previously used for dial feet. This indicates a change of movement

A mid eighteenth-century thirty-hour longcase clock by Charles Oldham of Southam which started its life as a single-handed clock (as is evident from the presence of quarter-hour markings, and the absence of minute markings). It is now fitted with a two-handed movement. Note that there is an unused hole for a dial foot just below the 12, confirming that changes have been made

An alternative method of making a single-handed movement into a two-handed one is to add another wheel to the train whilst retaining the same dial. Motion work has to be added to operate both hands in sequence. This results in the centre arbor for the minute hand being in a slightly different position. These changes demand relocation of the dial feet to fit the movement. At first sight the changes suggest a simple marriage of a dial with another movement, but a closer examination will clarify the situation. (Colour illustration facing page 160, lower.)

Additional plates have been attached to the dial plate of the clock by Charles Oldham to support the new dial feet and thereby allowing the modified yet original movement to be re-fitted. It is obvious that these changes have been in place for many years

A very common outcome of a change of dial is that the winding squares are no longer correctly positioned in their respective holes in the dial. When originally made the winding squares would have been perfectly centred. A departure from this suggests a replacement dial. However, as accumulated wear occurs the exact alignment of the winding holes can be lost. In weight-driven clocks the square will move downwards due to the pull of the weight. With spring-driven clocks it is not so clear-cut and depends upon the layout of the trains. Clocks provided with fusees would normally show displacement downward and to the side. This loss of alignment must not be confused with a poorly fitting dial but as an example of where some restoration is necessary. In the event of misuse it is possible for the winding squares to become bent or twisted. When this has happened it can be quite difficult to ascertain if the squares are correctly centred or not. If practicable, spinning or rotating the winding squares will give an estimation of the average position and a judgement can be made on the accuracy of alignment.

A movement from an eight-day longcase clock in which the winding square for the going train does not line up correctly with the hole in the dial, necessitating the removal of the dial ferrule. This is clear external evidence that a different movement has been fitted to the dial

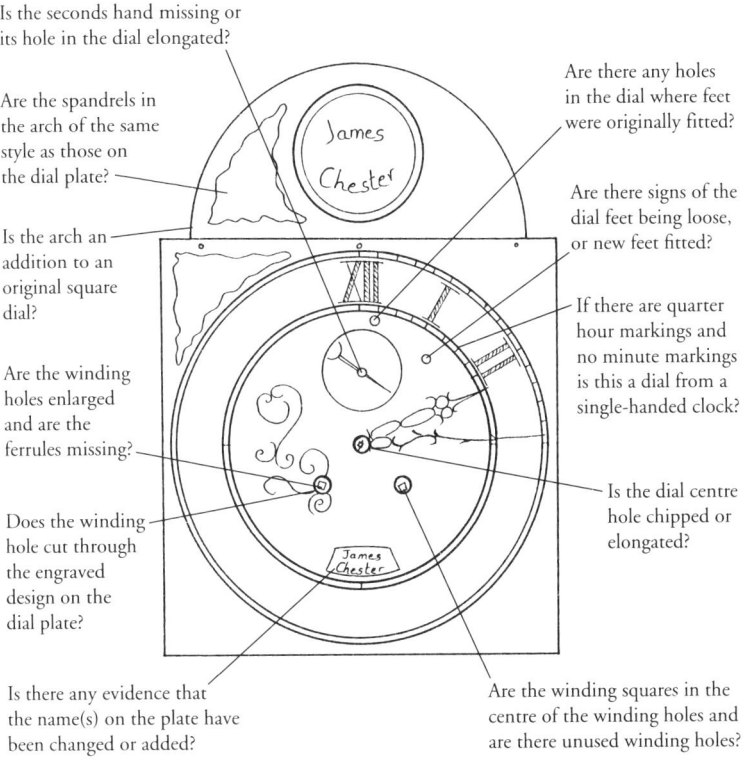

Is the seconds hand missing or its hole in the dial elongated?

Are the spandrels in the arch of the same style as those on the dial plate?

Is the arch an addition to an original square dial?

Are the winding holes enlarged and are the ferrules missing?

Does the winding hole cut through the engraved design on the dial plate?

Is there any evidence that the name(s) on the plate have been changed or added?

Are there any holes in the dial where feet were originally fitted?

Are there signs of the dial feet being loose, or new feet fitted?

If there are quarter hour markings and no minute markings is this a dial from a single-handed clock?

Is the dial centre hole chipped or elongated?

Are the winding squares in the centre of the winding holes and are there unused winding holes?

James Chester

Fig. 14   Possible external signs of a marriage of dial and movement

Clocks with French circular-plated movements have the enamel dial or chapter ring retained by two or three stout copper wires or pegs, soldered on prior to enamelling and bent over or pinned so as to be retained in an intermediate or false plate. This method of fixing allows a slight movement of the dial and helps to minimize any stressing of the enamel. New or unused holes will indicate that the dial has been changed. Alternatively, the winding holes may not line up correctly with the winding squares and may have led to damage to the enamel on winding. Care needs to be taken when making a judgement solely from damage around the winding holes. Slight movement of the dial can lead to the risk of damage on winding, and this is not uncommon.

Some of the late nineteenth- and early twentieth-century wall and mantel clocks have an enamel or plastic chapter ring with a separate inner ring for the winding holes. Changes to the inner ring are, therefore, much easier but rarely warranted in clocks of recent manufacture.

An eight-day dial will almost always have the seconds arbor vertically above the centre arbor. The winding squares are normally horizontal and equally spaced from the centre arbor. (There are departures from these general rules, but rarely does the dial look balanced and it was not a popular approach.) This layout places a constraint on the fitting of a replacement dial. It is not possible to rotate the dial when attempting to fit it, otherwise the winding squares will be out of position. A change in the vertical position of the winding holes becomes necessary unless of course the layout of the holes is identical or nearly so – which does happen. When it does, it is only necessary to relocate the feet. One very effective way of accomplishing this is to make and fit a complete false plate. An old unused plate may be used if suitable. Cast iron will rarely be available and sheet steel will have to be used. These changes are clearly seen from a close examination.

With some exceptions, movement replacement is easier with a thirty-hour dial. There is only one hole in the dial to consider, the centre arbor. It is possible therefore to rotate the dial without problem and this is done specifically to ensure that the feet engage with the front plate of the movement. If a clock is found where the bottom of the plates of the movement are not parallel with the bottom edge of the dial, a marriage has been made. Evidence of this type of change may be found on the case to accommodate a movement which is out of vertical. The baseboard will either have been cut away or set out of horizontal, both of which are easy to identify.

The bottom of the movement is not parallel with the base of the dial. This necessitates a seatboard which is not level. This is an unlikely occurrence in an unaltered case and strongly suggests that the movement does not belong with the dial

A method occasionally used to make a movement fit a dial is to replace the two plates of the movement completely whilst retaining all other component parts. This is an easy but time-consuming task. The purpose is to allow the layout of the arbors to be altered slightly. In this way all of the holes in the dial can be matched perfectly with the movement as well as the dial and false plate fixing. It leaves none of the normal evidence that a change has taken place, but it can be done successfully only with a suitable movement. A good-quality result can be achieved, but it clearly lacks originality. In a fully restored movement it can be difficult to spot. Lack of signs of wear or rebushing to rectify wear in an old clock are features to look for. The colour and finish of the modern brass plates may also give a sign that they are replacements. Modern rolled brass will have none of the flaws, holes or areas of porosity often found in cast brass and will be very even in thickness, due to it having been rolled rather than hammered – old hammered movement plates can have large variations in thickness. Modern cast sheets can be obtained for clock plates but brasses do vary in colour. The best way to identify replacement parts is to look for major differences in colour. (Colour illustration facing page 176, upper.)

To make a movement fit a dial another approach is to make partial changes to the layout of the train within the existing plates. In this way, for example, a seconds arbor can be moved to fit an existing hole in a dial, or the two winding squares moved to fit the two winding holes. New holes are required in the plates as well as the filling of the unused holes. This leaves distinctive evidence but it can be hard to identify without a thorough examination of the movement.

English dial clocks incorporated the traditional method of attaching the movement to the dial with or without a false plate. Marriages of movement and dial will show similar evidence to that described for bracket and longcase clocks. (Colour illustration facing page 161).

Are Any Parts of the Movement not Original?

Complete replacement of the movement or dial is not uncommon, but even more likely is that individual components are not original. This could arise because parts of the movement were unserviceable or lost.

Peripheral components are the most likely external items to be lost or damaged. These may have been replaced with equivalent parts of a similar age and type and any concern about originality will then be lessened.

Brass spandrels on a dial are sometimes partly or wholly replaced. This can be done to rectify damage or deterioration of the original items. Cast items are usually sound, although pieces may be missing from the corners. Repairs are common. Items stamped from thin sheet are very prone to deterioration and become very fragile, so it is worthwhile checking each one carefully for originality, including those in the arch of the dial. A change of a spandrel usually necessitates a new fixing hole in either the spandrel or the dial. It may be necessary to consult reference books to check whether the spandrels are of the correct type and size for the dial; clocks can be dated from the type of spandrels installed. Often the spandrels leave an impression on the dial plate showing the outline of the originals. Some modern spandrels will be roughly finished, whereas originals will often have the flashings (thin pieces of cast metal) on the edge of the castings filed off, giving a much clearer outline. However, less conscientious makers will have left the spandrels in a rough form and it is their condition which will be the best guide to age and originality.

Items such as the bell, pendulum, weights and key can be exchanged during the life of a clock. Each type of clock has its own style for each of these components, and they should have been replaced with ones of a similar style and age. This is often not possible due to the lack of availability of suitable items and, compromises will have had to be made. A check on other clocks of a similar type will soon show the type and style to be expected.

Component parts of the movement are less likely to be changed, and only when the originals are no longer serviceable. This is most likely to be due to the failure of a component to operate correctly, thereby causing other problems with the clock.

A dial plate showing clear evidence of the outline of the spandrels which helps to clarify whether those now fitted are the original ones. Two have been missing for a considerable period of time. The chapter ring shows slight traces of the original silvering

Components may be missing and exchanged in some early clocks for a special reason. Lantern clocks and many of the early bracket clocks incorporated a type of escapement – the verge – which utilized a short bob pendulum. An improved version was invented later – the anchor escapement – which employed a longer pendulum and significantly improved the accuracy of the clock. This was the motivation for many clocks having the escapement converted to the new system. To restore these clocks to their original form, it is common for them to have been reconverted, although the ethics of this is a matter of professional debate. Changes of escapement have been carried out in other types of clocks, which may fall into the category of items with a special interest.

Components may be missing for other reasons. The most common is that a component has failed and been removed. Alternatively, one aspect of the clock may have ceased to function correctly and may have been affecting the performance of the rest of the clock. This can involve removal of parts of the striking work, chiming work, date indicator, moon work, maintaining power and automata. In some movements, the whole of the striking train will have been removed. In this case, one of the winding squares will be missing but usually the barrel with its winding square will have been left, perhaps to disguise the loss of the missing components. The most common missing item is the driving pin or wheel to operate the date indicator. For this reason the date hand may be missing or fixed, or the aperture for a date indicator may be covered.

The operation levers for rolling moon dials, date wheels and automata are frequently missing because they have ceased to function properly. Additionally, the movement may have been installed in a different case and some components may not have fitted. This can happen with bells and gongs which are too large.

Where the dial has been detached from the movement it is easy for components of the motion work (which drives the hands) to be lost. This is usually clear, as some items will still be bright and clean from being covered by the now missing components.

It is clearly much better when all items are original. Major omissions are usually reflected in the price of the clock. Where items have been replaced with parts of a similar quality and style the value will be largely unaffected.

Maker's Name

It is worth making an examination to assess whether the name on the dial is that of the maker of the clock. It is not unknown for the name to be altered or one to be added where none was present. This is done to give the clock a false or improved provenance. A stolen clock with a new name may not be so easily recognized if it appears on the open market, and clearly a clock with the name of a famous maker will command a higher price. A clock from a well-known maker or a London maker should *always* be treated with caution.

The name of the maker should correspond with the style and quality of work which the maker produced. To be certain of this requires an in-depth knowledge, although for many clocks the information is not currently available. In addition, the style of the dial should correspond with the style which was in vogue when the stated maker was known to be in business. A guide to assist in identifying the maker is given in Chapter 7.

A clock with a brass dial will have the name engraved on the dial plate, or on an attached plate such as the chapter ring or a separate name plate. To replace the chapter ring is a major task but it is a fairly easy matter to replace the name plate. However, it is difficult for all but an expert to copy the style of the original engraving. If a replacement is suspected the first step is to look for signs that the engraving has been done by an engraving machine. These use a rotating cutter and the termination of the cuts will be rounded rather than the square terminations characteristic of hand engraving. If done by hand, like the original, look for differences between the old work and the new in the depth of the cuts, the smoothness of the curves and the termination of the letters, and signs of any errors in the work. The circles may have been cut on a lathe or using a guided cutter, and these will show no signs of hand work. If

An example of a maker's name, 'James Upjohn, London', on a plate attached to the dial, which is engraved in a different style and quality to the rest of the dial and which is likely to have been added or altered to give the clock an improved pedigree

the engraving gives no real clues it may be necessary to remove the name plate. A check can be made on the method of fixing, which may have been modified. The plate material can be examined. Differences in the colour and finish of the brass can be tell-tale signs, or there may be another name on the reverse.

Such changes can be difficult to identify if well executed. An easier change to see is when the original engraving has been removed and newly engraved over the top. This should be identifiable on an engraved chapter ring by a difference in level caused by the removal of metal. Alternatively, the name may have been removed, the plate hammered from the back to bring the front surface level again and a new name added.

The maker's name may be engraved on the backplate of the movement, with or without engraved decoration. The name should be consistent with the name on the dial. A name on the backplate would be hard to change, but it may be an addition; the backplate

An example of a fine quality clock with an original maker's name, 'Ellicott, London', on a plate with engraving in a similar style and quality to the rest of the dial

may have been engraved to enhance an otherwise plain clock and the opportunity taken to add a new name.

An unrestored early nineteenth-century English eight-day bracket clock movement with an engraved name on the back plate, 'Thomas Byworth, London', which should correspond with the name on the dial. Thomas was a member of the Clockmakers Company 1815–40

The false plate, when it is of cast iron, usually has the name of the dial-maker cast in, running horizontally along the top or the bottom. In some examples, the name on the dial is cast into the false plate and stamped onto the movement. Unfortunately, such good provenance is not common but when present is a useful sign of originality. It must be recognized that the name on the dial or movement may be that of the retailer or assembler who played a minor role in making the clock, having obtained the dial with false plate attached from one of the competing suppliers.

Painted dials may have been completely repainted, either by overpainting the original or by painting onto a stripped-off dial. When originally made there will be runs of paint on the rear around the perimeter. These should be present and suitably aged. Runs of fresh paint on the rear of the dial are a sign of a lack of originality, as is the absence of ageing due to handling or rust. Alternatively, the whole of the centre circular portion of the dial may have been repainted, or just the numerals and other markings and, when well restored, these will be an asset. It is easy for the name on a dial to be changed or lost during these processes.

Hands

An examination of the hands of a clock can provide useful information about originality. Both the length and style of the hands are characteristic of a particular type of clock and are very important in the visual appearance. When original, the style of the hands is a very useful guide in dating the clock. It will be necessary to consult reference books which give details of the date when each of the various styles were in use.

Early nineteenth-century painted longcase dial bearing the name 'W. Nicholas, Birmingham' with perfect correspondence between dial, false plate and movement since all bear the same name. See pp. 148 and 149

False plate attached to the dial is cast with the name of 'Nicholas, Birmingham'. Note the two unused upper holes in the false plate. These plates are clearly multi-purpose and were used for dials where no moon wheel was fitted. False plates such as this made of cast iron are brittle and pieces are often missing

The name 'W. Nicholas' is stamped on the front plate of the eight-day rack-striking movement confirming its originality with the dial and false plate. Note the porous area on the bottom right-hand side of the cast brass front plate of the movement

The length of each of the hands needs to be examined. The minute hand should always extend to reach the minute markings so that the minutes can be easily and accurately read. It can also extend slightly beyond. Where two circles are provided to frame the minute markings, the hand should not extend outside the outer circle, or if it does, by only a small distance. A hand which does not reach the minute markings is very suspect; it may have been broken and repaired, or be a replacement.

It seems to have always been the fashion for the hour hand to be broader than the minute hand. This allows for easy discrimination. The length of the hour hand is important and less well defined than that of the minute hand. First, it has to look balanced in comparison with the minute hand to assist discrimination. Secondly, it should reach the hour markings. Normally the hand will cover about 10–20 per cent of the length of the hour numeral when measured from the inner edge, but up to 40 per cent is also found. If an hour ring with quarter-hour markings is provided the hand will reach to the markings in the same way as the minute hand.

Repaired hour hands are quite common even though, unless it is a single-handed clock, it is not used for setting to time. The hour and minute hands, if original, should be made of the same material, e.g. steel, brass or copper. The profile, decoration, and finish of the hour hand should be either of a similar style to the minute hand or of a complementary style. Reference books will help to make this perfectly clear.

A seconds hand usually meets the same length requirements as the minute hand. It is sometimes in a different material from the main hands, although the style will generally be complementary to them. A date hand will often match other small hands on the dial, such as the seconds hand.

Deviations from these general rules may suggest a change from originality. Fortunately, replacement of missing hands is not too difficult

An early nineteenth-century longcase clock by Haynes of Stamford with matching brass hands which are the correct length for the minute and hour markings on the dial and with a brass seconds hand in a sympathetic style. Note the central fitting of the dial in the dial frame and the winding squares which are perfectly centred in their respective dial ferrules. Both factors suggest that the movement and dial are original to the clock. The aperture for the hood door lock is in the normal position in the lower piece of the dial mask

A late nineteenth-century balloon style clock with a circular plated French movement and dial which is of the correct size. It fills the bezel perfectly and helps to confirm its genuineness

and good reproductions are available. Neither should be of serious concern to the owner. What is most important is that the visual appearance of the clock is not impaired.

Carriage Clocks

Carriage clocks usually have the movement retained by two screws through the metal base of the case. The base is covered by a thin metal cover which is easily removed to access the securing bolts. The cover will often have been lost. Signs of unused or change holes in the base may indicate that the movement has been replaced. A more likely occurrence is that the escapement mounted on the platform has been damaged and the platform assembly has been exchanged for either a modern replacement or one of a similar age. It is rare that the fixing holes for a replacement platform align exactly with the original. Evidence may therefore be left of such a change in the form of elongated fixing holes or missing fixing screws.

If the dial is of card it could be a replacement for the original enamelled one.

A carrying case is often provided with a carriage clock. If it is original the clock will be a snug fit so that it is prevented from sliding around inside the case. A space is provided for the key. These cases are frequently in a distressed condition, but they are still an asset when original.

Clocks with Complications

A small proportion of clocks included a number of extra features, adding considerably to the complexity of the movement and dial. These are usually the finest clocks and many are in public ownership.

Clearly the better the quality of the clock and the higher the financial value, the greater is the likelihood that it has been kept safe and sound since manufacture. It is also more likely that its provenance is known over previous generations and that the restoration is of the highest standard. Nevertheless, a good provenance does not guarantee this. It is even more important when considering the purchase of such items that professional advice is sought.

Example of a platform escapement, with lever-type escapement (lever is arrowed), mounted on the top of the movement of an English carriage clock. A missing screw on the top left-hand side of the platform, as a consequence of misalignment of the fixing holes, is the sign of a replacement platform

An early nineteenth-century eight-day longcase clock incorporating an attractive extra feature – an Adam and Eve automaton in which Adam offers his apple at every alternate tick of the clock. Note the hands (particularly the minute hand) are short of the markings and may well be replacements

5 Assessing the Quality of a Clock

An assessment of the quality of an antique or collectable clock will be of benefit in making an estimate of its value and, if it is being offered for sale, whether it is a suitable item for purchase. It is useful as a part of judging the extra expense of restoration or repair.

It is tempting to make an overall assessment of quality based upon the attractiveness of the finished result. This must generally be resisted since tastes in fashion determine the models in the market place and current views can only be subjective. It is better to use the objective matters which are outlined here. Nevertheless, highly attractive clocks are usually of good quality; the skill involved in producing a high-quality clock is complementary to the skill required for producing attractive designs.

It has already been suggested that a clock by a well-known or top maker is likely to be of good or even excellent quality. Although this is usually true, the prospective owner ought to be in a position to be able to make an independent judgement. A clock which has no maker's name, or one with a hitherto unknown name but which is of excellent quality, may be found in a shop or saleroom. Alternatively, the clock may have had the name of a famous maker added at a later date, or be from a family of makers who used the same name, although only one member produced top-quality work. The following advice is intended to provide a basic and objective guide to assessing quality, irrespective of whether the maker was eminent or not. Of course, if the quality is judged to be high and there is an eminent maker's name on the dial, then the owner can be reasonably confident that the work is from the maker's workshop.

Wide variations were seen in the quality of clock movements and cases when these were made by local craftsmen to their own and their customers' tastes and designs. When mass production began this changed. After the middle of the nineteenth century wall, shelf and mantel clocks were produced in large numbers to the same standards and designs. Nevertheless, an assessment of the quality of these clocks is just as important.

It is essential when attempting to assess a clock's quality for the owner or prospective owner to be able to make a detailed examination of it. However, if the movement is not in a clean condition, it may not be possible to see some of the tell-tale signs.

The quality of a clock is based on four criteria:

- the materials used
- the design
- the amount of work involved in producing it
- the skill of the maker and others involved in its construction

The Case

Materials

The quality of materials used in constructing a clock case are reflected in how it has stood the test of time and use. Primary consideration will be given to cases made of wood, this being the most common material used. Other materials used in cases will also be briefly considered.

The type of wood used will be dependent upon the age of the case and the woods available and in fashion at the time. For guidance in this area, the owner should consult one of the many reference works on antique furniture. It is sufficient to state that the higher-quality cases will use the more expensive imported woods such as mahogany, ebony and rosewood. Oak was the primary material used before these imports became available and continued to be so in country areas where it was in plentiful supply. The best oak for cabinet work is termed quarter cut; these are slices cut between the centre and the edge of the log to give the narrowest gaps between the growth rings. It has the appearance of a series of fine and close straight lines. Cut in this way the shrinkage is minimal, since it is mainly the softer material between the rings which contracts. When used in furniture, fine-grained quarter-cut oak will have kept its size and shape well. It was used for the long side panels and doors of a clock case, where the opportunity for distortion was greatest. Coarse-grained oak, cut across the log, would have distorted. The same is true of the heart wood which is particularly prone to rot and infestation. A timber containing knots will be at risk of warping, and will have been avoided when choosing the best planks for construction. Where other types of wood have been used for the main structural components these same features will be evident.

The quality of the materials is an important factor in the overall external appearance. The best fine straight-grained timber will have been chosen for the long front and side panels, with the grain running in line with the longest dimension. This will have achieved the best appearance, with maximum strength. No knots should be found in these pieces. A pronounced grain would be used only to give a pleasing and balanced effect.

There is less to say about the quality of metals or other non-timber materials used in cases. Metals should have no holes or imperfections. Marbles would be expected to be flawless and have attractive figuring in a high-quality clock. Glass panels would be flat and clear, with the minimum of bubbles or flaws.

Design

The primary design consideration is to ensure structural strength and stability. Methods of construction followed changes in the availability of new types of wood. Details of these changes can again be obtained from authoritative sources on the construction of furniture. A good design employed strong joints and ensured that the changes in dimensions resulting from drying were adequately compensated for. Any wide panels needed either to have fine-grained straight timber or be allowed to move to prevent cracking. When fine-grained hardwoods became available these were used in combination with oak or with a carcass of oak or a softer wood such as pine.

Veneering was used as a way of employing the higher-cost materials in the minimum quantities. A secondary benefit was the enhanced strength achieved when two materials, which could move in different ways, are attached together. A good structural design with a veneered carcass had adequate strength and good resistance from warping. Softwoods were used for carcasses when a lower weight was required. Where veneers have been used, the best pieces will be cut from larger sheets to give a good visual effect and these will be of an attractive style. Matching pairs of veneer will have been chosen carefully to give a symmetrical and pleasing result in good work. If the overall effect is that the case looks unbalanced, then a compromise will have been made in choosing the materials, and this is unlikely to have warranted the best work in subsequent fitting and finishing. A piece of veneer which has had a patch inserted to make good a fault in the original piece will not be seen in good work. This will have been left for a lower-grade piece.

Evidence in an early nineteenth-century longcase dial of loose feet causing paint to be lost. The dial has correctly matching but non-original hands; the minute hand is clearly too long for the minute markings. The seconds hand is also missing. These external features point correctly, in this example, to a different movement having been fitted later

The movement attached to the clock by Charles Oldham of Southam shows unused holes in the front plate of the thirty-hour movement suggesting that it had previously been fitted to a different dial. In fact, changes have been made to the layout of the going train of the movement to allow for two-handed operation and it has then been re-fitted to the original dial. The porous nature of the cast brass is evident on the front plate

A good example of a completely original fitting of an eight-day fusee movement to the dial in an English office dial clock. Note that a sheet steel falseplate is used

Opposite page and above: A late nineteenth-century trunk dial clock show-ing 'ears' which have become detached and one completely lost. The extent and quality of the carving makes an otherwise ordinary clock into a very attractive and ornate piece. Note the absence of the winding square for the striking train. In this clock the striking train is completely missing

The external proportions of a case are the second aspect of design. They must ensure adequate strength whilst giving a balanced and attractive result. The proportions were largely set by the fashions of the day. London styles generally led provincial and country fashions by many years. Deviations from the generally accepted proportions would have been unlikely to find a buyer. For pendulum clocks a slender trunk or body with a more prominent dial or dial surround has generally been the prevailing fashion. For spring-driven clocks a more uniform width of case is normal. Gross forms are generally avoided and when they were produced, additional steps will usually have been taken to give the impression of a slimmer style.

The third aspect of design is the use of detail. A complicated design with copious detail can lead to a significantly more attractive case and a high-quality result. Cases with decorative floral or geometrical inlays prove this conclusively. The use of features such as compound or carved mouldings, carved, fluted or reeded pillars, canted corners, pierced friezes, intricate inlays using various coloured woods or other materials and cross-banding all add a charm and interest to an otherwise plain shape. Just as important is the appropriate use of attached decorative items. Handles, finials, ormolu mounts, cast frets, gilt mouldings, fancy cornices etc. add interest which makes a plain case look – and be – expensive. The number of items attached will be a guide to the quality, as will the amount of detail on each piece of applied decoration, on a sliding scale, with the most detailed, and therefore the most expensive to produce, being used on the best-quality cases. For example; cast brass capitals on the pillars on hoods and cases will be Corinthian on the best cases and Doric on the lower grades; on mantel and bracket clocks bezels to retain the glass will be heavily cast with attractive designs in quality work.

The apparent proportions of a case can be altered by the appropriate use of decoration. For example, a large case may be made to look narrower and lighter in appearance, as was done with the larger longcase clocks. However, the excessive use of added decoration can easily result in an over-elaborate and heavy overall appearance which defeats the original objective. The correct level of decoration is clearly an important aspect of good design. (Colour illustration facing page 176, lower.)

A bracket clock bearing the name 'Upjohn, London' in an ebonized case. The case has additional external fittings reflecting the extra work and cost in such a piece

Other features of good design which will be found in good quality work are:

- veneers which meet in regular geometrical shapes or are mitred to give a balanced appearance
- the consistent use of stringing throughout the whole of the case, base and hood
- the sensitive use of different types of coloured veneer
- the use of pictorial motifs as shells and Britannia; this may not necessarily be on the best work – they are more likely to be used to create a good impression!
- the use of decoration to enhance strength or stability, such as the cross-banding of the edges of doors and panels
- the copying of features in a dummy form to give a balanced appearance, such as pairs of matching escutcheons
- the use of carvings or cast mounts which are complex and crisp, giving an impression of depth
- a good likeness or representation in the use of animal or natural forms

One matter which may need to be considered when assessing a high-quality piece is that the design will have been specially commissioned by a client with a particular location in mind. This may have required aspects of the design to fit in with the other furnishings and fittings of a room. Such a piece could have been destined for a stately home or a palace, and its design may not conform with other clocks of the same period.

An inevitable consequence of a complicated or detailed design is the requirement for a greater amount of work to accomplish it. A higher level of skill will also be necessary and it has to be well executed if it aspires to be quality work.

An example of a well made late eighteenth-century oak case, which displays close attention to detail. The runners which keep the hood in position are slanted downwards towards the back with an angled under-surface, ensuring that as the hood is pushed into position it is retained both precisely and tightly – and it still is after 200 years

The Amount of Work Involved

The amount of work involved will primarily be determined by the size of the clock case, its design and its complexity. The greater the number of additional items which are present and the more extensive the decoration, the greater is the work required. Some of the finest clocks are also the tallest but this is not a factor which can be used to judge quality in isolation.

The second factor which determines the amount of work is the standard of finish on each of the components. The better the standard of finish, the better the quality. Ormolu and cast brass fittings should be without blemishes such as holes from the casting process, and cleanly finished to reveal the design clearly with no rough internal or external edges. Gilding will be even in colour, except where different colours are deliberately used to highlight features such as foliage. Surfaces of veneer should have been rubbed down to a completely flat surface, although the movement of wood and shrinkage with age can cause problems in this area. Additional work in finishing those normally hidden components to a high standard is done only for the most discerning of clients, is a sign of good attention to detail, and is usually associated with high quality.

Skill

A skilled craftsman will be able to achieve the required standard of work in a shorter time and execute it with fewer minor errors. The level of skill is impossible to judge from the end result, since more time may have been taken to achieve the same outcome. Nevertheless, factors to look for are the quality and clarity of the finish on components. Hand carving should be crisp and clean, with repeated shapes all of the same size and proportions. Features will be deeply and confidently cut, with smooth and neatly blended curves. No filling will have been needed to make good any wood lifted when cutting into the grain. Where a component has errors in its preparation, a skilled person will discard it and make a replacement.

Where components are to fit together the accuracy and smoothness of the fit are a good guide to the skill of the worker. Examples are the fitting and removal of a hood onto a trunk and that of the door into its aperture.

Above and overleaf: An early nineteenth-century Indo-Gothic style English bracket clock case with a considerable amount of extra detail: cluster columns, turned and carved finials, decorative frieze, sunk panel with gilt mouldings and carving. It is evident that considerable additional work from a skilled craftsman is needed to produce a quality case. There is no maker's name on either the dial or the movement

167

Deeply cut mouldings with fretted frieze, finely carved and turned finials and finely detailed brass capital and bezel are all signs of a quality item

Another test of good quality is the number of minor flaws present. There will undoubtedly be some, but they should not be obvious without detailed examination. Whilst the quantity of carved or inlaid detail is not in itself a sign of quality work, only the most skilled will take it on and the more complex the carving the more confident and skilled the worker. With veneers, the absence of minor gaps between adjoining pieces is a sign of good quality. Where stringing is used, the lines should be of even width, and should be straight when they are intended to be. Shading of veneers to simulate shadows will be uniformly balanced in tone and extent.

Perfect fitting of cluster columns and evenly spaced carved decoration on feet and mouldings all require extra effort and skill to produce a quality result

The Movement and Dial

Materials

Brass for the movement and components of antique clocks was cast and then hammered to provide a hard, flat surface. Good-quality brass will be of an even colour, with no signs of cracking or porosity. Porosity will appear in poorer-quality sheet brass as a series of small holes or small particles which were not fully melted together in the making of the brass. The best pieces will have been used for the movement plates and components. Sheet brass containing flaws will have been used in places where it cannot be seen, such as on the rear of dial plates. Deep pitting in the surface of decorative brass castings suggest poorer-quality work, preventing a good finish being obtained. In mass-produced clocks, the brass was machine rolled and will give no valuable clues; the only guide to quality is resistance to wear.

Steel will have been forged to the overall outline and then hand finished, leaving no external evidence, on well-finished parts, of the material grain. The best way to judge the quality of materials employed is by their condition after considerable service, assuming that there is no evidence of severe neglect or misuse. Any porosity in the surface of the brass for the wheels can lead to the failure of items under stress such as teeth and, under the application of ammonia-based cleaning agents, any porous brass in wheels, components or plates can become brittle and weak.

Design

Mass-produced clocks were designed with thin plates, skeletonized to minimize the use of metal. As a consequence, wear is usually rapid and lack of rigidity only accelerates this failing. Steel components will often have been stamped out from a sheet. The cut edges will be left entirely unfinished in low-grade work but will be well finished in the best.

The design details for movements such as those in longcase clocks were widely known, as is evidenced by the remarkable similarity of movements over a considerable period. The best work will usually employ thicker material for key components such as the plates, or an extra pillar to support the plates, giving extra rigidity.

Weights, which are a utilitarian aspect of a clock, will probably be cased in brass for quality clocks, even when these are enclosed in the case and are not normally seen. A pendulum bob may have graduated scales, both on the bob and on the case, to allow easy and accurate adjustment on the best pieces. Decorative cast brass items will often have deeply cast designs.

Assessing the Quality of a Clock

Although the early bracket clocks were individually made, later movements were made by specific suppliers to different types and sizes, and are generally of a high quality. The use of a steel-linked chain in spring-driven clocks, rather than gut, implies a higher quality due to the extra cost.

The addition of extra, non-essential features will make any clock more useful. Indicators of the phases of the moon or the state of the tides, automata, zodiacal periods, indicators of the state of winding (up and down), a choice of chimes, rate adjustment, sun time etc., were all included for those prepared to purchase a 'top of the range' model for private or business use. Most of these extra features usually showed on the dial, giving a more attractive and desirable product.

The style of a dial was very much a feature of popular fashion and the changes in fashion over considerable periods have been researched in detail.

A most important aspect of a clock is the clarity of the dial, which determines the ability to read the time quickly and accurately in an age when the level of illumination in the home was much lower than today. (Colour illustration facing page 177, upper.) This was fully appreciated by the best makers and a failure to meet this requirement suggests a lack of quality and taste. Very many recently made clocks suffer badly from this feature. Any engraved decoration, at its best, will be fine and balanced, avoiding crude heavy features which would reduce the clarity of the hands when viewed against the backdrop of the dial.

Where a new feature was invented by a maker, such as the system of rack striking to allow repeating of the last hour, temperature compensation to the pendulum to provide more accurate time-keeping, or maintaining power to maintain going during winding, these would have been employed initially in the best-quality work. The discerning owner was prepared to pay extra to be up to date with the new technology.

The use of jewels in a movement to reduce friction was not widely employed. This is restricted to regulator, high-quality or special clocks.

The Amount of Work Involved

The surface finish on cast brass plates in good-quality work will have been obtained by polishing, after hammering, filing and scraping had produced a flat surface with an even plate thickness. In addition to looking better, this had the practical benefit that dust tends to accumulate less on a polished surface and oil is less likely to drain away from where it is required. Where there is a visible grain left on a component, this will be arranged to lie along the major axis in an attractive way. In some of the best-quality work, the plates will have a repeated pattern scratched over the surfaces. Lower-quality work will often have the brass plates left in the hammered condition with no attempt to provide a perfect surface. The evidence of marking out for the wheels and other components will usually be left untouched. Edges on plates and all brass components will be filed square and straight on good work and on best-quality work will be crisp. On cast spandrels and other decorative items, all imperfections from casting, such as flashings, will have been removed on good work and the surfaces polished. Steel components will be filed and polished on good-quality work, with no signs of file marks, even on hidden surfaces. Screws will have neatly shaped heads, consistent in shape and style throughout the clock, depending upon their purpose, and will be either highly polished or blued. The rounding of edges was regarded as poor work, being easier to produce, unless the rounding is for other reasons such as enhanced mechanical strength. Chamfering of corners, if used, will be neat and of a uniform width.

In the best work, in addition to the actual work involved in making of the clock, consideration will have been given to its subsequent repair. Components will be identified, usually with small punch marks of a unique number or pattern, on each pair of mating components. This is desirable where there are a number of similar components such as springs and screws, and is very useful for the restorer. The purpose is so that on assembly each item is uniquely identified and can be put together in exactly the same way. This is common on good French movements and on top English work. Another method of marking of components is for each item to be marked with the same number. This would have been useful to aid identification when a batch was made. It is not in itself a guide to good quality, just an indication of the close control in manufacture which the best makers will have used.

Another sign of quality work is where simple functional components have been given attractive shapes, e.g. levers with scrolled ends or minor changes so that parts resemble animal or natural forms, which all require additional work in shaping and finishing. Brass wheels will have been 'crossed out' in almost all good-quality clocks to remove the bulk of the mass. The finer and lighter the end result, whilst maintaining the necessary strength, the better the quality of the work.

Where the key is original to the clock this can be expected to be constructed and decorated to a similar standard as the clock itself.

The amount of work needed to produce a dial, and hence its quality, will be determined by the number of additional items which are required to show on the dial and on the level of decoration. Engraving, such as in the centre of dials and around the perimeter, matting in the centre of dial plates, piercing and engine turning are techniques which were frequently used to give a more attractive effect. None of these is in any way essential for the operation of the clock and they require considerable time and skill to execute, usually carried out by an outworker or itinerant craftsman to the

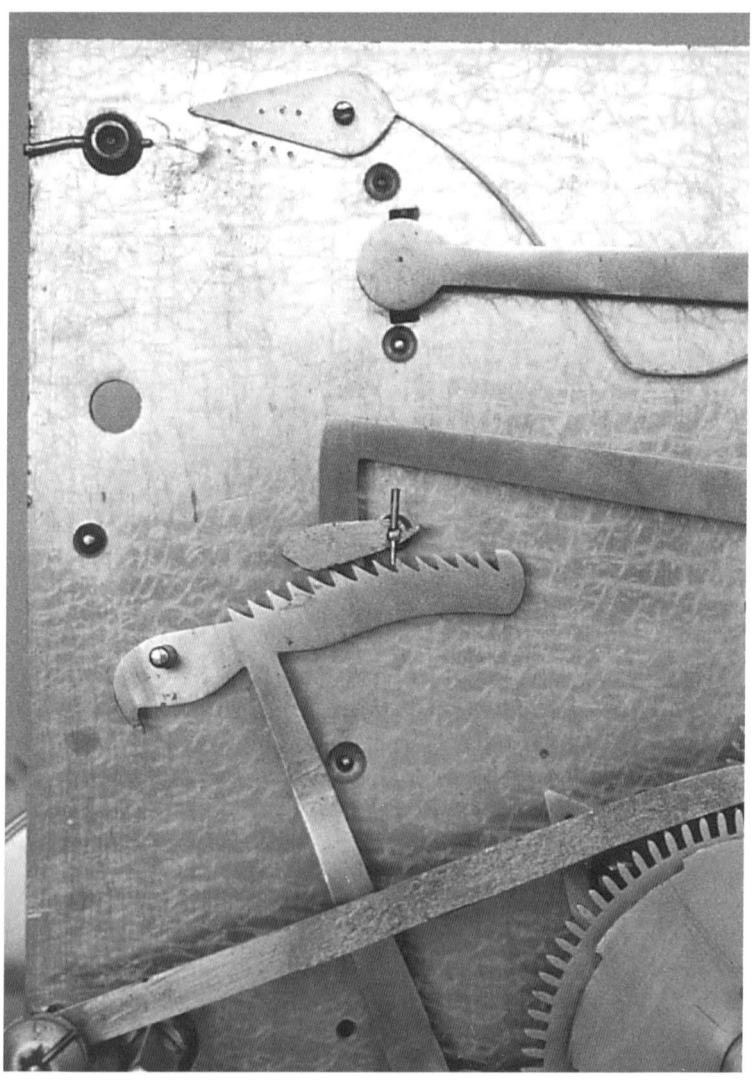

An example of high quality work in an eight-day striking and chiming long-case clock by O. Brandreth of Middlewich where the components are provided with identifying punch marks so that springs etc. are correctly installed. The front plate is covered with freehand scratched marking to provide an attractive result

In the high quality work in this eight-day striking and chiming clock by Ellicott the plates are square and very well finished. All other components are completed to a good standard. Chiming is on eight matched bells with the hours striking on a gong.

design of the maker. It is not always possible to judge the quality of the horological work by the quality of the decoration, but its presence indicates that the maker was seeking to produce a high-quality clock. Decoration can be found around the edge of the dial plate, often in the form of herringbone engraving, around the date aperture and on the chapter ring and subsidiary dials. The backplate of the movement will be used as a location for decoration and identification in better work. Similar decoration on the pendulum may be included.

In the manufacture of enamelled dials, a number from each batch will have been lost in the firing and handling processes. This will have been due to the difficulties in producing perfect surfaces with accurate markings in a medium which is made molten to fix the

The rear of a brass dial showing the hammered surface and flaws in the cast brass which are not visible from the front. Usually, the brass plate is cut away behind the chapter ring to avoid waste but in this high-quality clock by Ellicott it has been left *in situ*. There are no visible signs of changes to the surface. Note the strike-silent components which are fully operable

Part of the door and trunk of an early nineteenth-century Yorkshire longcase with reeded columns and inlaid cross-banding and both pictorial and geometrical inlays. The cabinet work has been executed with good quality materials and an attractive design which involved considerable extra work. Restoration of the boxwood stringing to the left of Britannia has not met the original standard

A very good quality brass dial, eight-day striking and chiming clock by Ellicott of London, a well known family of makers of fine clocks. The photograph illustrates the high clarity of the dial, the fine quality gilt spandrels and the very neat engraving

An example of an unrestored painted dial with very little evidence of the original black lettering below the date aperture but which retains an impression of the maker's name and origin (Robert Walker of Montrose). Sometimes strong reflected light, or ultraviolet light, is necessary to see the markings. This dial will be fairly easy to restore

design. Enamelled dials are therefore normally associated with a higher quality than painted ones. Failures in manufacture will have been more frequent when dials had additional items such as subsidiary dials, so enamelled dials with additional features are the sign of a better-quality piece.

Above and opposite: French mantel clock in Belgian black marble case with red marble inlays. A good example of quality and perfect originality. The minute markings on the dial are correctly positioned within the bezel confirming the originality of the dial. The winding squares are perfectly centred in their holes. Note the square above the 12 o'clock for fine adjustment of the rate, the substantial cast bezel and the thick bevelled glass

Painted dials, sometimes termed 'white dials' or 'japanned dials', were made to the requirements of the clock-maker with different suppliers offering a range of styles. Each extra feature had to be specified and each incurred an additional cost. Large dials cost more than small ones.

Skill

When making and finishing the components of a clock, the maker will have planned the design, layout and finish of each component. A quality finish cannot be achieved unless it is planned into the making of each component. The presence of unfinished areas is a tell-tale sign. The best makers will have aimed for a standard of finish which is high and uniform over all items. Lower quality will have a higher standard of finish only on those areas in view. In judging these matters, the observer will have to ignore items which have been the subject of damage, repair or replacement, although the best-quality restoration work should have achieved a similar standard to the original work.

The type of finish will also depend upon the age of the item. Where items are polished, the quality of the polish is a useful guide, but a shiny finish is not in itself a guide to quality. Obtaining a highly polished finish whilst retaining the crispness of the edges of components is an important skill and will have been mastered by the best workers. Craftsmen who took a pride in their work and recognized the importance of attention to detail are those who are likely to have produced the best. These are the same people who are likely to have included in the work their own marks of identification.

Although the quality of the engraving done by an outworker on a brass dial is not a direct indication of the skill of the clock-maker, a maker setting a high standard in his own work will have

French circular plated movement with count wheel (locking plate) striking. These French movements are usually of good quality and restore well. They often have the facility to adjust the rate using a small key from the front of the dial, as is visible at the top of the movement in this example

Engraved and painted moon dial in the arched dial of the clock by
Obadiah Brandreth (see opposite)

been prepared to pay for a high standard in the work done for him.
The standard of the engraving can be judged by both the design and
the cutting. Good, consistent, shaping of lettering and numerals and
the correct spacing between the letters are essential for quality work.
Animal and natural forms will be correctly modelled. Appropriate use
of shading is important, as is attention to detail in the form of serifs
and terminations. Quality cutting is evident from the smoothness of
the curves and evenness of grading between different widths and
depths of cut. Straight lines will show no deviations. There should be
no visible errors where the tool has slipped or cut outside the
intended area. The absence of these is a good sign. When metal has
been unintentionally cut out, even though it has been burnished in an
attempt to hide it, this removes only the sharpness of the cut. It is not
possible to replace the metal. After a serious mishap, minor amend-
ments to the design are necessary to hide the error and a slight lack
of symmetry of a design may then be found.

Opposite and above: Mid eighteenth-century longcase clock by Obadiah Brandreth of Middlewich. The good quality of the engraving of the dial is evident. The letters are neatly spaced, the script is flowing and the numerals finely drawn. The external features – the moon dial, strike/silent and date hand – all add up to a high quality clock

Parts of an eighteenth-century brass dial showing insufficient attention to detail. The strike/silent dial has uneven and poor-quality punching and the engraved numerals on the chapter ring are badly formed

The dial from a late eighteenth-century Yorkshire thirty-hour longcase clock with attractively designed engraved decoration, well executed with good shading to give a pleasing result. However, there are numerous differences in the two matching scroll designs and the proportions of the bird are poor suggesting a dial and hence a clock of average quality

Well executed engraving on a mid eighteenth-century 'Willis of Harthill' brass thirty-hour dial, with an attractive herring-bone engraving around the edge. However, there is evidence of bad planning in that the name overlaps with the numerals

Competing individuals and companies sold painted dials, each offering their own styles. The quality of the finished dials varied greatly. Some painting is of a very high standard, often where portraits are featured as a part of the design. The quality of the decorative painting is usually the best indication of the quality of the dial. In the period before coloured scenes were used, gilt spandrel motifs extended over the corners and into the arch of a dial, with gesso to give the design some relief, and they are often extremely attractive. Some of the worst-quality dials were produced

A thirty-hour painted dial from a clock by John and Edward Mason of Worcester. An example of a lower quality dial. The base paint is very thin and the decoration poorly painted

in the last half of the nineteenth century. These utilized a thin base paint layer with limited, often simplistic, decoration and are frequently found in poor condition.

It is likely that aspects of a painted dial will have been altered or restored and this may make the assessment of its quality problematic. A good restorer will frequently be able to rediscover the original style to allow the true quality to be assessed.

6 Assessing the Condition of a Clock

The prospective owner needs to make an assessment of the condition of a clock. It may be possible to call upon the services of a horological expert to offer advice but often this is not feasible and it will be necessary to make an independent assessment. It may be desirable to make a rapid critique and judgement if an opportunity to buy is not to be lost.

The following advice covers the most important aspects, and does not require expert knowledge, just a careful examination of the key parts of the clock. Because there are so many different types of clock, each of which has been modified in some way over its life, it is impossible to cover all eventualities, but there are many common features, which makes it easier to assess the condition of a particular clock.

The Case

Those who are familiar with antique furniture of any kind will have no difficulty extending their experience to clock cases. Damp and woodworm are two of the worst dangers to wooden items, causing rot, warping, loss of veneers and mouldings and loss of surface finish. Evidence of their presence is usually fairly obvious.

Darkening or discolouring of the wood, water marks and decay are all signs of dampness.

Woodworm is commonly found in old cases and should not in itself deter the prospective purchaser. Holes in the surface of the wood and fine loose dust are clear signs of woodworm. Active worm is usually associated with fresh dust, and can easily be treated. It is only a serious problem when the wood has lost its strength so that it crumbles under load or is unsightly. When present in modest amounts, the surface holes can easily be filled in with a coloured wax on visible areas. When present in major amounts the wood will have to be cut out and replaced, although if it is in the carcass of a veneered case, it may be impracticable to effect a replacement.

The other primary concern is previous restoration. When viewing a clock which outwardly appears to be in good condition, there are items that the potential purchaser needs to be wary of. Many cases offered by dealers today have been completely stripped and repolished. Sometimes the standard is excellent, but sometimes it is either poorly or over done. Whilst this may help to sell the clock to certain purchasers, over-restoration is not good practice. If it is done because the surface is in very poor condition, then it is acceptable. Otherwise, it is taking away much of the originality. It is much better to refinish areas which are distressed in a style to match the good remaining parts.

When an assembly of wooden components is made up into a case, each item will have been made to fit correctly, but shrinkage will occur to a different degree in each of the items and this may lead to distortion of the overall shape. This can result in the hood looking rather twisted. It may have caused cracking at one or more points along the grain on panels which are restrained at their edges. This may be found in the front panel of the base of a longcase clock, in the top of a bracket clock case, and less frequently, in the sides of hoods, trunks and plinths. Severe warping can be very difficult to counteract and could become more pronounced when a case is moved from a cool, damp area and kept in a warm, dry room. It is best not to try to reverse warping unless it is causing other problems. In severe examples, partial rebuilding may be the only solution. When it is less severe it may have to be tolerated and gaps filled with wood, wax or other fillers. Such action is rarely fully convincing and a clear sign of the originality of the case is lost.

Missing components such as mouldings may have become detached because of distortion due to warping or excessive heat, or due to dampness, which causes glues to soften. Remaking and fitting mouldings to a suitable colour and finish is a skilled process for a restorer and may involve considerable expense if a number are missing.

Veneer is often lost from parts of a case, and this can be fairly easily restored. The use of old pieces of wood of a similar age to the original case give by far the best result. Only when the veneer is missing in quantity is it a matter of concern to the buyer. The veneer can be affected when the underlying wood has shrunk to a different degree, leading to rippling or cracking. When this has occurred it is hard to rectify, requiring professional restoration. Stabilization of the underlying wood may be required.

Cases may have the original surface coated with one or more layers of varnish. With time the varnish will usually have deteriorated to leave a streaky surface, and have turned to an unattractive dark brown colour. When viewing a case coated with such a layer it is useful to scrape a small area of the case with a fingernail or coin somewhere not too obvious. This should reveal the underlying surface. When the coating is completely removed the original wood surface, often in a well-protected condition, can be cleaned and properly finished. In auction rooms this fingernail test will often have already been done by a professional buyer or restorer. If there is evidence of strong staining of the wood then more care is needed to be certain that the case will respond to restoration.

Another type of coating which can hide the original wood surface is a thick layer of wax. When applied over many years it can accumulate dust and dirt. The end result can be a hard, dark layer. This can be removed fairly easily, usually revealing a well-preserved surface underneath.

Guidance on how to identify the presence of recently applied chemical stains was given in Chapter 4. Once stain has been applied it seeps into the grain of the wood, particularly into the end grain, and is difficult to remove completely. Where it is seen to be present, it may be necessary to judge whether it is a detriment to the overall appearance and condition.

Water and rust stains appear as dark, intense marks on the wood, often in the shape of the offending item. Burn marks are sometimes seen on cases, appearing as a brown mark, or as a depression where the charred wood has been removed. These types of blemish are to be expected in a piece of old furniture and are a matter of personal taste or dislike. Restoration may be necessary.

Cast items or inlaid strips of brass may be missing or damaged. Close examination of all cast brass items for signs of stress cracking or crumbling is recommended because they can cause weakness and necessitate replacement. Suitable modern replicas can usually be obtained at reasonable cost. Inlaid brass can be restored, but at a high cost, especially where parts of an intricate inlay are missing. Brass bezels should be examined for signs of deterioration. Cast bezels are usually very sound, although the hinges may be bent or loose. Thin pressed metal bezels are more vulnerable to damage and deterioration.

Glass panels or domes which are cracked or chipped can be unsightly, as well as allowing the ingress of dust, and may warrant replacement. Flat glasses are cheap to replace but are not a good substitute for an original bevelled glass, which considerably enhances the appearance of a clock. Panels with bevelled edges, particularly those which are shaped, and domes can be expensive to replace – only a limited range of sizes of domes is available. Chipped or cracked glass panels in a carriage clock or other glazed cases can be evidence that the clock has been dropped and extra care needs to be taken in examining the condition of the movement.

Metal cases will suffer bending or denting if dropped and will fracture if made of cast metal. Repairs should have been in keeping with the original style and manner of construction.

Marble and slate cases should be carefully examined for signs of chips off corners and cracked panels which a restorer will be hard pressed to disguise. These cases are usually cemented and wired together, but they may have lost some of their adhesion and strength. Occasionally, a panel may have been replaced to repair damage. The surface colour of black marble cases is often distressed but can be restored to a good standard. Coloured marbles will require more specialized restoration.

Ceramic and china cases often suffer minor chips and cracks. If repair is required it is specialized and expensive. Unless the damage is unsightly or the mechanical strength of the case is reduced, a repair is probably not of benefit.

The Movement

When a clock appears to be in good or reasonable condition, the potential owner will first wish to find out whether the clock is operating correctly or is in a potentially going condition. In a shop there is every chance to be shown all aspects of the operation of the clock, including how to bring the chiming back into correct sequence should it become necessary. It is clearly in the seller's interest to be able to demonstrate a going condition in every respect. In an auction room or when a clock is offered 'as seen', the buyer will need to make an assessment. This is not always easy but there are a number of basic checks which can be made which will go a long way towards covering the main potential problems. Where the examination leads to severe doubts expert advice can then be considered.

The first check to make is on the general state of the movement. If it is dirty, rusty or smothered in oil it is unlikely to respond to the tests suggested below. Nevertheless, it is worth proceeding because any clues will be useful. The second step is to ascertain that the weights, pendulum, key, bell(s) or gong(s) are present.

One of the most important reason for these checks is to determine whether all of the operating parts of the clock are *in situ*. If any are missing, repair or restoration will require considerable extra effort.

It must be strongly emphasized that a clock is a delicate mechanism and the checks detailed below must be carried out with appropriate care to avoid any risk of damage.

Spring-driven Clocks

The clock will usually be in a partially wound condition. Assuming that this is so, first check the striking and chiming. The first step is to move the minute hand carefully forward, pausing after each quarter to allow the clock to strike or chime. *The hand must not be forced if resistance is felt.* If no striking occurs at the hour, or chiming at the half or quarter, it is necessary to apply the key to the winding squares for the striking (at the front, usually the one on the left-hand side) and for the chiming where fitted (usually on the right-hand side), to test whether the clock is wound. In an auction room the key and pendulum will not usually be with the clock, as they are often retained in a separate place for safekeeping, but staff will be happy to provide them and may assist with trying out the clock. After checking that the key fits correctly it should be wound by a half or full turn, if possible. Again, if there is resistance, force must *not* be applied. A second attempt can then be made to move the hands forward. Success in chiming or a partial movement of the hammer is a sign that the train components are intact. The bell or gong may be missing and no chiming may be heard, but the lifting of the hammer as well as a quiet mechanical action are adequate signs. Lack of activity is not proof of a problem, possibly just a sign maintenance is needed. If the key turns freely, with no resistance, then the chances are that the spring is broken and a new one is needed. If the key winds the spring, without the familiar click of the winding ratchet, and on releasing, springs back, there is a failure in the winding ratchet. Both of these shortcomings will need the services of an expert, but neither is usually too serious.

Next check the going. If the pendulum is in position, it is necessary to give it a slight swing and listen carefully for signs of ticking. If none is heard, the pendulum should be carefully moved from side to side by hand, with an increasing arc at each oscillation, again listening for evidence of ticking. If the pendulum is not present, do the same action to the crutch into which the pendulum fits. Ticking is a good sign and indicates that all components are present. Lack of a tick may be due to the spring being wound down. It will be necessary to apply the key, wind a little if possible and try again.

A clock with a balance wheel instead of a pendulum needs to be given a quick twist whilst holding it firmly in the hand. This should make the balance wheel oscillate freely. If it does this is a good sign, whereas if it slumps over to one side a major repair will be necessary. It is desirable to allow the clock to run for sufficient time to ascertain that the escapement is operating, allowing the train wheels and the hands to move forward. If they do, the going train is complete and is probably in reasonable condition. The hands are driven by the clock through a simple friction device. If they are loose they may not be driven. In this situation the performance of the clock should be judged by the correct operation of the escapement and train.

A spring-driven clock which has been correctly restored should operate freely with no more than one full turn of winding. If it has to be nearly fully wound to give satisfactory operation, it is likely that additional maintenance or repair will be required.

Clocks with other features such as a 'strike/silent' lever, date hand, four/eight bell selection etc. can be carefully moved by hand to make them operate. It is quite likely in an unrestored clock that these extra features will not operate. Nevertheless, it is advisable to check if the lever is operating the mechanism, even if the clock does not perform the function so as to ascertain whether the essential components for operation are present.

If the clock passes all of these tests, it should give confidence as to its completeness. The better the clock fulfils these requirements, the better the condition of the movement is likely to be. General cleaning and maintenance may still be required and advice on how to assess the needs in this area is given below.

Weight-driven Clocks

If the weights are fitted in their correct positions, similar tests to those above can be carried out. If not, it is necessary to apply power by pulling on the rope, gut or chain. This will only be worthwhile if the gut or chain is sitting correctly on the barrels. It should not be attempted until the rope or gut is correctly installed. Alternatively, it may be necessary to apply pressure to the large driving wheel or barrel onto which the gut or chain is wound. This technique of applying pressure to the driving barrel will have to be used for a spring-driven clock where the mainspring is broken. A little trial and error will be necessary to find which of the cords to pull or which direction to apply pressure. An equivalent force to the normal weight will be required. Pulling on the wrong cord will cause no harm to the movement, but pressure should *not* be applied to any of the other wheels or arbors in the clock.

Other Movement Checks

When a movement has been seen to be clean and has been demonstrated to be in operating condition, three additional checks need to be made:

1 The pivot holes in the rear of the movement plate need to be inspected for either clean oil or dirty, congealed oil. Clean oil suggests recent restoration. Dirty oil or the complete absence of oil implies that cleaning and correct lubrication is required.

2 The backplate needs to be examined for the presence of a coating of oil on all or part of its surface. Excessive oil is not good for a clock and cleaning is probably necessary.

3 The pivot holes in the backplate need a close examination to determine whether they are elongated due to wear in the plate. Any evidence of this indicates that maintenance is required. The more severe it is the more urgent the repair.

When a movement is not in clean condition, another eight checks are necessary.

1 Evidence should be sought of rust on all steel surfaces. Minor traces suggest that maintenance is required. Ingrained rust on the pivots and on other steel components suggests that serious deterioration has occurred. Some degree of surface

rusting is to be expected in the oldest movements and can be tolerated. Many parts will still operate, although they look unsightly, but others will not and replacement will be necessary. Expert advice must be taken where evidence of recent rusting has occurred as the movement may be past economic repair.

2 A movement subjected to fire will probably have suffered terminal damage. Extensive rust, blackening of brass items and blistering of paint are signs of fire damage. Softening of the hardened components will have rendered them useless.

3 If the brass plates of the movement are bent or distorted the movement can usually be restored, provided that the wheels and other components have not been bent or damaged.

4 A movement completely covered in grime can be almost perfect underneath. It can hide a rusted movement totally beyond repair or even hide the evidence of exposure to fire. It requires close inspection.

5 The heavy construction of antique clocks is such that they normally restore well provided that any rust is not severe. French cylinder-type movements used in mantel and carriage clocks usually restore to a high standard.

6 Where components of a clock are missing then the cost of restoration will be considerably higher.

7 Where a clock has a balance wheel as the timekeeper instead of a pendulum, it is usually mounted on a small plate on the top or back of the clock, hence the name 'platform escapement'. This is commonly found in carriage, balloon and some mantel clocks. If the clock has been knocked or dropped it is this delicate escapement which tends to suffer first. If the balance wheel does not swing freely and the escapement does not operate, the chances are that repair will be needed, and is usually expensive. There are two primary

types of platform escapements: lever escapements and cylinder escapements. All types are normally jewelled. The lever type is often recognizable from the short steel lever which projects from near the escape wheel. In some examples, the lever is nearly hidden underneath the balance wheel. In a cylinder type, no such lever is present. The easiest way to identify a cylinder type is from the fact that the edge of the escape wheel on the platform comes very close to the centre of the balance wheel. The cylinder types are more vulnerable to damage and less easy to repair. The lever types are less prone to wear and are better timekeepers. Replacement platforms are sometimes fitted and should be of a similar style to the original if they are not to devalue the clock. Modern platforms are of the lever type. A number of other types of escapement were fitted to carriage clocks, the identification of which will require specialist knowledge.

8 It is always a good sign when a clock is complete with key, pendulum, weights etc. When one or more of these are missing it is unlikely to have been in regular or recent use.

The Dial

Brass Dials

A brass dial should be intact, with no evidence of splits or cracks in the front surface. These are very difficult to rectify. A dial which is bent can be straightened provided the bend is not so severe that a crease is left behind. If there are deep marks, gouges or scratches it may be impossible to remove these and if they are severe they will always leave an unsightly reminder. Frequent polishing may have

A detachable platform on which the escapement is mounted is found in some mantel clocks and in carriage clocks allowing the clock to be portable

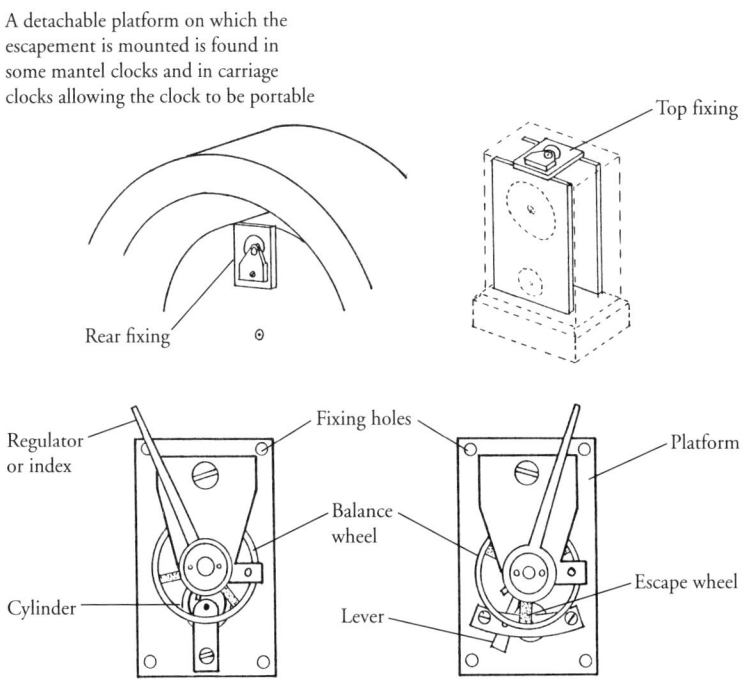

Top fixing

Rear fixing

Regulator or index

Fixing holes

Balance wheel

Platform

Cylinder

Lever

Escape wheel

CYLINDER ESCAPEMENT TYPE

The escapement consists solely of the balance wheel and the 'cylinder' escape wheel. This type is more prone to wear and damage

LEVER ESCAPEMENT TYPE

The escapement consists of a balance wheel, escape wheel and a lever. This type is more desirable, easier to repair and a better timekeeper. The lever is not always obvious

Various other types of escapement are found on platforms but the cylinder and lever are the most common

Fig. 15   Platform escapements and how to identify them

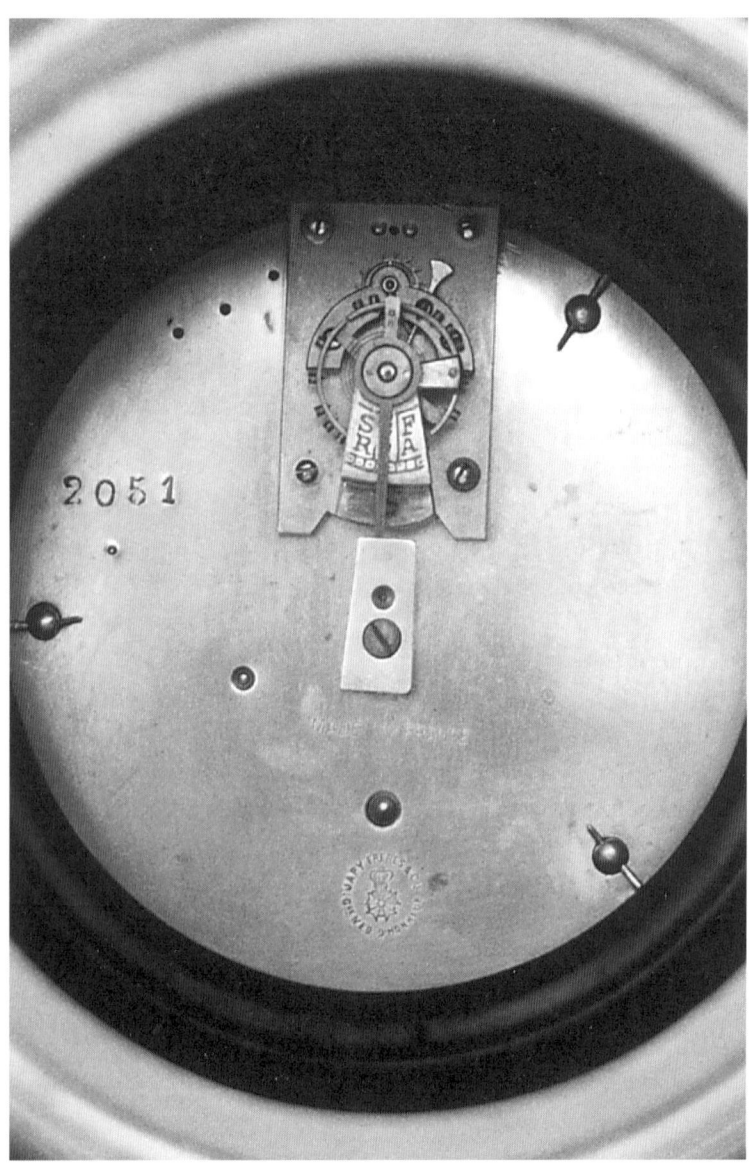

An original platform escapement fitted to the back plate of a French balloon-style clock. The small silver lever on the top right-hand side of the platform indicates that this is a lever-type escapement

worn away details of the engraving and rounded the edges of the remainder. This is restorable but will require a skilled engraver to ensure that it matches the original work and could still result in an undesirable loss of originality.

Where the engraving is filled with a black wax, part of it may be missing. It is an easy matter to restore this. The chapter ring and other subsidiary dials and attached plates were almost always silvered and then lacquered. Traces of the silvering are often visible even after a dial has been polished. Silvering is easy to restore and makes a large difference to the attractiveness of the dial. One difficulty with resilvering is that some dials have deep pitting where corrosion has occurred. It is not possible to remove these pits, and when the dial is resilvered they may show as unsightly dark marks.

The centre of the dial is often engraved or finely matted. Any damage is hard to restore without some trace being left behind.

Painted Dials

Painted dials can suffer from a number of problems due to vandalism, neglect or natural deterioration. Some of these are serious and some are minor.

It is the centre portion of the dial which suffers as the owner moves the hands or winds the clock. A majority of painted dials, therefore, have the decorative paintwork intact but with the numerals and name either worn or overpainted. Reapplication of the numerals and circles is fairly easy to achieve and can make a considerable improvement to the overall appearance of the dial. Where the numerals are badly overpainted and unsightly, a satisfactory restoration will require the overpainting to be removed prior to remarking. Overpainting of the whole of a dial is not uncommon and is usually evident from the poor standard of painting. It will have to be removed with care, but it is usually successful and will frequently reveal many aspects of the original design. This is work for a professional restorer if further damage to the dial is to be avoided.

It is quite common to find that paint is either lost from the surface or is loose and easily detached. This can occur with the main dial and particularly with other dials such as the moon and date dials. Loss of paint can arise due to rust, fire damage, acid attack or delamination due to the unsuitability of the base layers of paint. When the dial plate has started to rust, the corrosion can extend a long way underneath the paint layer, causing it to buckle off where the rust is most pronounced. To effect restoration it is essential to strip off all rust and loose paint from the affected areas. A flat surface of the correct colour has then to be built up. Only

An unrestored dial from an eighteenth-century English eight-day longcase clock showing the deterioration in the numerals and other markings through prolonged use. Note the enlarged winding and centre holes indicating a modification of the dial to fit a different movement.

then can restoration of the artwork and numerals begin. This is a time-consuming process and is strictly for a professional restorer. As a part of this process, the original artwork will be covered by a layer of protective varnish so that in principle the dial could be returned to its cleaned, but unrestored, condition. Only where none of the original surface can be saved will a complete repaint be necessary.

If a dial has been bent or dented, this will probably have caused some of the paint layer to become loose or detached. If the dial is straightened, additional paint will come off. Where the bending is

A considerable improvement has been made to the overall clarity and appearance of the dial by having only the numerals, circles and minute markings correctly restored without resorting to extensive repainting and thereby detracting from the originality of the dial

minor it is probably best to leave it. Where small holes have been drilled in a dial these can be filled and restored.

Prior to restoration, it is necessary to remove all dirt and oil. In the case of oil, the dial will require degreasing and may leave staining. Once the dial is clean, the restorer will use an ultraviolet lamp to seek details of the original numerals and markings on the dial. This will be valuable in restoring the dial to its original style and condition.

Deterioration of the painted decoration or pictures in the corners, arch and centre ground is difficult to restore to a good standard. For this reason it is often left unrestored. The artwork on painted dials can be of a very high quality and often utilizes materials such as gesso and gold or silver leaf. As a consequence, restoration should be done to the same high standard. Provided that the paintwork on a dial is all intact good restoration is possible at reasonable cost. Where the work required is more substantial, it may be advisable to obtain advice on the likely restoration costs prior to purchase. (Colour illustration facing page 177, lower.)

Enamelled Dials

Many enamelled dials have fine cracks due to the in-built stresses in the dial or to carelessness in fitting and use. These cracks can be cleaned but will darken again as grime enters. Dials which are cracked or broken cannot be perfectly repaired. Touching up with enamel paint can improve matters, provided that a good match to the original colour is achieved, and this is a cheap method. Professional repairs can also be done, which give a good result but at a high cost. The more complex the markings on the dial, the greater the cost to repair.

When considering purchasing a clock with a damaged enamelled dial, it is wise to weigh up the matter carefully. Those with perfect dials have a considerable premium.

An early twentieth-century oak striking mantel clock in which the dial shows evidence of severe damage around the winding holes. However, this is not due to any change but is a consequence of the loose fitting of the dial and bezel allowing the dial to interfere with the key during winding

General

In addition to the above guidelines, there are some very broad generalizations which should give the intending buyer a better 'feel' for the item. They will not cover all situations, but they will often be found to be true.

1 If the clock has been lovingly cared for then this is obvious from the general condition, and will also be reflected in the price. These items usually appear at good auctions and in private sales.

2 It is easy to store clocks such as bracket, carriage or mantel clocks in a dry condition over long periods without major deterioration; their small size makes it more likely that they have been kept in good condition. This is probably why bracket clocks are not often found in a completely neglected condition, although they do appear with modifications or parts missing and occasionally in non-original cases.

3 When they are not operating or required, longcase clocks are more likely to be stored in a shed, barn or loft, which is often entirely unsuitable for the case, dial and movement. As a consequence, they can appear for sale in a disreputable condition. It is quite common to find a clock which has been completely neglected on which, to obtain a reasonable price, the seller has done some superficial work to make it look better. This could well be to its detriment. An untouched patina on the case can often be revitalized with care. Once it has been taken off, it will take many more generations to restore it. Any rough overpainting of the dial may be difficult to remove. Any abrasive cleaning of the movement will be irreversible. A completely unrestored clock is often a better buy because the owner will not need to rectify any inappropriate repairs.

4 Many longcase clocks have movements and dials which are not original. This is less true of other types of clocks.

5 A neglected clock which is enveloped in grime inside and out has probably not been subjected to recent repair or bad repair but it is probably in need of some mechanical maintenance.

Fakes

Having made a judgement about the condition of a clock and its originality, the owner or prospective owner may still be left in doubt about its authenticity. It is possible that it is a complete fake, made in recent times. The movement, the case or both can be faked.

Movements can be copied but this is a rare occurrence. The work required is not normally warranted by the potential value of the end result. However, early lantern-type clocks are fairly simple mechanically and of high value, and fakes are found. In fact, it can be quite difficult to tell a fake or a later copy from the real thing where old materials have been used in the construction. This is an area best left to experts.

It is much more likely that a wooden case is faked, since this can be accomplished more easily and can be better disguised. The fake case may house an old or a modern movement. A modern maker or restorer can produce pieces of furniture or clock cases which are hard to distinguish from old items. The prospective purchaser must know what specific aspects to look for so as to be able to discriminate between them.

A fake case is usually revealed by a careful internal examination. There should be no evidence of recent stain or treatments on the carcass woods. Ideally the wood should be bare, with only the signs of knocks and bruises accumulated over the years. There may be a little absorbed oil from the movement. When making a modern piece the faker will have to make all the wood look old. It is easy to do this on the external surface, using stain covered by varnishes and polishes, and where appropriate, coloured or gilt lacquers, and every part of the external surface will be covered. It is only on the inside surfaces where the faker will have a problem. It will be necessary to use stains, varnishes and polishes to hide the new wood. Stain, when applied to new wood, will penetrate to a different extent depending on the grain: on the end grain the stain will penetrate deeply, while on the side grain it will penetrate to different degrees, sometimes leaving a streaky finish or a run. Varnish or polish on the inside of a case should be treated with suspicion, except where it has clear signs of antiquity.

A good test for newly stained wood is to make a small cut into it to reveal its internal colour, although on the end grain the stain will have penetrated deeply and testing can be misleading. Having cut through the layer of stain the brightness of new wood will shine through. Deep cuts may not be practicable but there should be no problem in making one at a hidden point inside the case. Try to find an area where there is already a small knock on a corner, where the new wood may show through. With old wood the colour will be the same all the way through.

Cracks in the wood can provide additional information. In an old piece the wood will have shrunk and small – and sometimes large – gaps will be present, particularly when a panel is wide. If it is possible to see into the cracks the wood should have the dark colour expected from mature wood.

One aspect of an old case which should be immediately apparent is that the wood is very dry, particularly when it has been indoors for a prolonged period. This aspect is not easy to replicate. In a case made from fresh new wood, the wood will appear to be more dense, with a closed grain, and will feel cooler and moister to the touch.

Mention must be made of cases made at a significantly later date than the movement but which are now of considerable age. These do appear in the market place from time to time. It is only by gaining a knowledge of similar cases and the movements which they contain that these marriages can be identified for what they are.

Finally, a case may be primarily of recent manufacture but utilizing some small part of an original case or piece of antique furniture.

For this reason, a judgement on a case must be made from a general examination and not just on selected parts.

7 Identifying the Maker of a Clock

It has been the custom from the earliest times for clock makers to make markings of some kind to identify their work. This would normally be on the dial or on one of the plates of the movement. For those belonging to the Clockmakers' Company, it was a requirement that work was correctly identified. Marking therefore shows a compliance with professional membership as well as a clear pride in the work done and a strong sense of ownership. In the majority of clocks, these and other types of markings are still present and are a very useful starting point in finding out more about the maker.

The first step is to examine the dial and plates of the movement carefully for markings, being always on guard for possible changes to the originality. Markings may include the name of the maker, family of makers or company, usually with the town or village of residence, and sometimes with the address.

On a painted dial this can be above, below, or level with the centre arbor. The lettering will usually, but not always, be horizontal. It is also common around an arc of a circle. Arched dials can bear the name in or around the perimeter of the arch. Although the paint may have totally disappeared, or just have left a pale shadow behind, by examining it with the dial at an angle to bright light, some evidence of the original lettering can often be seen. Alternatively, ultraviolet light can be useful in picking up evidence of the remaining specks within the light background.

On an enamelled dial the lettering may be fused into the white enamel or added later in some form of paint or black varnish onto the surface. In this latter case, it is easily removed and often lost. In special dials, the lettering may be scratched into the surface of the enamel. Where separate enamelled panels are used for each numeral, there may be an individual plate solely for the maker's name. On brass dials, the name can be engraved onto the dial plate or onto an additional plate. A wide variety of different positions on the dial or in the arch have been used. The position was often a part of the prevailing style at the time of manufacture.

On the plates of the movement markings are most commonly found on the rear of the backplate (the surface facing the back of the case). The markings may be scratched, engraved or stamped. Markings on the front plate are less common. Circular plated French movements will usually have a stamped mark on the backplate and a works number. Good-quality bracket clocks may have an engraved backplate and this will often include the name of the maker. On plain backplates scratched marks are sometimes found giving the name of the maker with the date. Unfortunately, this occurs on less than one in twenty longcase clock movements. False plates between the movement and the dial, and bells, sometimes carry the name of the movement maker but it is much more common for these items to bear the names of the foundry or dial maker. Mass-produced wall and mantel clocks will usually have a manufacturing mark stamped onto the plates.

Makers' markings are sometimes found inside the case door, on the backboard of the case or on the baseboard supporting the movement. In addition, there may be markings by restorers or previous owners. Restorers' marks are usually on the plates of the movement and less commonly on the rear of the dial or on the pendulum bob. Markings are sometimes in chronological order. Until recently, it

Original maker's marking on the rear of an unrestored eighteenth-century thirty-hour longcase movement which may help in proving the authenticity of the name on the dial

was common practice for a restorer to mark a piece upon completion. This is a practice which is not now recommended since the making of marks on antique clocks is considered as defacing the item.

Identifying the Maker of a Clock

Once a maker can be attributed to a clock it is then possible to obtain information about him or her from published sources. Research has been carried out by a number of authors and many reference books have been published giving basic details of makers in a majority of the counties within the UK, as well as in Europe and elsewhere. These books usually give makers' details and the family relationship to other makers, together with their places of residence. The primary sources of this information are trade directories and parish records. Information on clocks and watches known to have been executed by them, as well as those in museums, is usually included. Where it is known that a maker was working in a particular period or at a specific date, these dates are normally given. Knowing these dates it is possible to compare the style of the particular clock with the styles in vogue at that time. This can be most helpful when there is more than one maker of the same name.

Local museums and restorers will often be able to offer advice regarding local makers and their work. They may also have a number of sources of reference.

It can be of considerable interest to the owner to have information regarding the maker and history of the clock. An additional benefit is that amongst some owners and potential owners there is a desire to have a clock with a clear provenance, particularly where the clock is of high value. Inevitably, this can lead to a differentiation in the value of the items in the event of sale.

Bibliography

Clock and Watch Makers

There are two books which combine many of the local records and cover the whole of the world.

Baillie, G.H., *Watch and Clock Makers of the World*, Volume 1 (NAG Press, 1988)

Loomes, Brian, *Watch and Clock Makers of the World*, Volume 2 (NAG Press, 1989)

There are numerous other books which give additional details of makers in particular geographical areas.

Clock Styles, Cases and Movements

There are many books available covering most types of clocks. Some of the more useful in each area are listed. Omission of a book from the list does not imply a disregard of a particular source of reference in terms of its quality or coverage. The list is intended to provide a taster for the enquiring owner who will be able to make a start on an enjoyable road of discovery.

Allix, Charles, *Carriage Clocks: Their History and Development* (Antique Collectors' Club, 1981)

Bibliography

Dawson, P.G., Drover, C.B. and Parkes, D.W., *Early English Clocks: A Discussion of Domestic Clocks up to the Beginning of the 18th Century* (Antique Collectors' Club, 1982)

Loomes, Brian, *Complete British Clocks* (David and Charles, 1978)

Loomes, Brian, *White Dial Clocks: The Complete Guide* (David and Charles, 1981)

Ortenburger, R., *Vienna Regulators and Factory Clocks* (Schiffer, USA, 1990)

Rose, R.E. *English Dial Clocks* (Antique Collectors' Club, 1978)

Royer-Collard, F.B., *Skeleton Clocks* (NAG Press, 1994)

Tennant, M.F., *Longcase Painted Dials: Their History and Restoration* (NAG Press, 1995)

Thorpe, Nicolas M., *The French Marble Clock* (NAG Press, 1990)

Information for Owners

Barker, D., *The Arthur Negus Guide to English Clocks* (Hamlyn, 1980)

Bird, Anthony, *English House Clocks: An Historical Survey and Guide for Collectors and Dealers, 1600–1850* (David and Charles, 1973)

Good, Richard, *Keeping Time: Collecting and Caring for Clocks* (British Museums Press, 1985)

Loomes, Brian, *Antique British Clocks: A Buyer's Guide* (Robert Hale, 1991)

Loomes, Brian, *British Clocks Illustrated* (Robert Hale, 1992)

Smith, E.P., *Clocks: A Guide for Owners* (Lutterworth Press, 1993)

Price Guides

Curtis, T., *Lyle Price Guide to Clocks and Watches* (Lyle Publications, published annually)

Miller's Antique Clock Pocket Guide (Mitchell Beazley)
Shenton, A and R., *Collectable Clocks: Reference and Price Guide, 1840–1940* (Antique Collectors' Club, 1997)

Restoration

Wills, Peter B., *The Conservation of Clocks and Watches* (British Horological Institute, 1995)

Museums

Many provincial or university museums contain clocks and watches as part of their collections, generally specializing in local makers or industries. There are good collections in Leicester, Lincoln, Birmingham and Cambridge, to name but a few.

The principal collections in the UK are at the following museums:

Upton Hall, The British Horological Institute, Upton, Newark, Notts
Manor House Museum, Bury St Edmunds, Suffolk
The National Maritime Museum, Greenwich
The British Museum, London
Museum of the History of Science, Oxford
National Museum of Science and Industry, London
Prescot Museum, Prescot, Lancs
The Museum of London, London
Museum of the Worshipful Company of Clockmakers, London

Index

Index

Index

219

The French Marble Clock
Nicolas M. Thorpe

French Marble clocks are increasing in popularity at a time when many antique clocks are becoming too expensive for the average collector. Popular with the Victorians, they are now enjoying a well-deserved revival. Most have high quality movements with attractive cases made in a wide variety of styles, and fortunately there are still many of these clocks to be found at reasonable prices in antique shops, junk shops and market stalls.

The French Marble Clock is the first complete book on the subject, written by Nicolas Thorpe who is an enthusiastic collector. Years of research coupled with many visits to France and Belgium have made him a leading authority about them. The opening chapters reveal the history of the marble clock along with new information about their origins and makers. Photographs of factories, now demolished, are set alongside material from French archives.

To many collectors the most valued part of the book will be the practical chapters. These take a detailed look at the famous *pendule de Paris* movement that is found in many French mantel clocks from the 19th and early 20th centuries. Sound and sensible advice is given about examining, dating and buying, together with a clear explanation of the function of parts of the movement. This is essential reading for those wishing to restore timepieces or clocks with striking movements. The chapters on restoration and clock case styles are all clearly illustrated.

One section on Collector's Clocks shows some of the world's finest marble clocks, and one appendix has an invaluable list of known French, German, American and English makers and their trade marks.

In his Foreword, Michael Turner, of Sotheby's, London, describes these clocks as often being 'ridiculously underpriced'. This carefully researched study of French marble clocks will be a valuable guide to all collectors whether they buy clocks in fine order or enjoy the challenge of restoring a damaged bargain to full working condition.

243 x 178 mm   288 pp   b&w illus. throughout
0 7198 0230 X

Longcase Painted Dials
Their History and Restoration

M.F. Tennant

Here, for the first time, is a book dealing exclusively with the history and restoration methods of the longcase painted dial.

The result of nearly twenty years concentrated restoration work by the author, this book has a generous selection of illustrations from her personal photographic library. It covers one hundred years (1770–1870) of a stylistically unique form of art originally developed in Britain, including examples which range from the truly sublime to the utterly ridiculous. Many of the clocks have been lovingly cared for and passed down through several generations. Others have not been so fortunate, suffering from the inevitable deterioration resulting from neglect and/or ignorance.

The practical section on restoration includes much sought-after and invaluable advice on dial problems plus a discussion of the differing methods and types of materials used. This book is essential reading for anyone with a longcase painted dial and for all those interested in the restoration and history of this fascinating art.

234 x 156 mm   256 pp   163 b&w & 24 col. illus.
0 7198 0260 1

Antique British Clocks
A Buyer's Guide

Brian Loomes

This book is the only one of its kind to deal extensively with the buying and selling of antique British clocks, revealing commercial knowledge acquired by the author over twenty-five years. Loomes concentrates on important background information for the novice, including distinguishing between the genuine and the fake, and between the good and the mediocre, as well as how to recognize those factors which represent quality. The guide examines aspects that influence price, where to buy (auctions or dealers), researching makers and styles, and 'investment' ideas, as well as restoration principles.

Despite the wealth of facts and experience recounted in the book, its style is accessible, and Loomes anticipates and answers the novice's questions at a comprehensible level.

Illustrated in black-and-white, *Antique British Clocks* will help the buyer choose successfully from the vast variety of clocks available.

234 x 156 mm   224 pp   70 b&w illus.   0 7090 4611 1

British Clocks Illustrated

Brian Loomes

The first fully pictorial survey of clocks and their history, styles, periods, qualities and collectability. Each page is self-contained, with each photograph carrying a full explanatory and descriptive caption so that salient points can be seen at a glance.

The book is divided into sections, covering topics such as lanterns, brackets, hooded clocks and longcases. With large, high quality illustrations, and many photographs taken specially to reveal points of details, *British Clocks Illustrated* forms a useful reference for restorers as well as taking a uniquely practical approach to its subject.

Surveying antique British clocks in historical/developmental sequence, Brian Loomes' book assumes no prior knowledge of the subject, and will prove invaluable to collectors, restorers and enthusiasts at every level.

234 x 156 mm   272 pp   300 b&w & 4 line illus.
0 7090 4547 6